PHYSICS RESEARCH AND TECHNOLOGY

RECENT DEVELOPMENTS IN DARK MATTER RESEARCH

PHYSICS RESEARCH AND TECHNOLOGY

PHYSICS RESEARCH AND TECHNOLOGY

RECENT DEVELOPMENTS IN DARK MATTER RESEARCH

NORI KINJO
AND
AKIRA NAKAJIMA
EDITORS

nova publishers
New York

For permission to use material from this book please contact us:
Telephone 631-231-7269; Fax 631-231-8175
Web Site: http://www.novapublishers.com

NOTICE TO THE READER

Additional color graphics may be available in the e-book version of this book.

LIBRARY OF CONGRESS CATALOGING-IN-PUBLICATION DATA

Recent developments in dark matter research / Editors: Nori Kinjo and Akira Nakajima.
pages cm -- (Physics research and technology)
Includes index.
ISBN: 978-1-62948-010-7 (hardcover)
1. Dark matter (Astronomy) 2. Dark energy (Astronomy) I. Kinjo, Nori, 1971- editor. II. Nakajima, Akira, 1973- editor. III. Series: Physics research and technology.
QB791.3.R43 2013
523.1'126--dc23
2013033804

Published by Nova Science Publishers, Inc. † New York

CONTENTS

PREFACE

In this book, the authors present current research in the study of dark matter. Topics discussed in this compilation include the submicroscopic viewpoint on gravitation, cosmology, dark energy and dark matter, and the first data of inerton astronomy; some new astronomical consequences of the dark matter conception; dark matter/energy, quantum mechanics and modern causal analysis; superheated liquids and the search for astroparticle dark matter; behavior of dark matter in interacting dark energy models; rotation curves in Bose-Einstein condensate dark matter halos; weyl dark fluid from the brane world scenario; massive neutron stars to dark matter particle properties; and modeling dark matter halos via solitonic BE condensates.

The so-called Standard Model of particle physics and the Standard Model of cosmology have so far not been able to suggest unambiguously what the very nature of dark matter is. These Models are therefore flawed at the fundamental level. In Chapter 1, the authors propose a deeper fundamental approach, the submicroscopic deterministic concept that considers the real space as a substrate constituted of primary topological balls arranging a tessellation lattice. Dark energy is accounted for by a peculiar distribution of topological balls inside the universe considered as a huge-range cluster in the total tessel-lattice. A moving particle interacts with the space generating spatial excitations named inertons by the author. Inertons being a substructure of the matter waves are present in any material object owing to the oscillating motion of its entities. Inertons form a vast series of harmonics that in the form of standing spherical waves spread far beyond the object inducing its Newton's gravitational potential. Starting from first submicroscopic principles the submicroscopic concept explained all phenomena predicted by general relativity. Dark matter is derived as a viscosity of space induced by standing inerton waves in the surrounding of star clusters. Two devices that measure inerton signals, which were designed by the authors'team, are described. Some measurements of the Earth and Sun inerton signals conducted by using these devices are demonstrated. Finally, the authors suggest designing the first inerton measuring complex, an inerton observatory, which will be able to measure amplitude, spectral and time characteristics of non-stationary inerton signals received from outer space.

Chapter 2 is dedicated to completely considering the new model of the halo theoretical description of dark matter, including new consequences of dark matter influence on the electromagnetic wave propagation – light rays and Cerenkov radiation.

In its introduction the authors have also briefly considered a few other astronomical consequences of the existence of dark matter as discussed in modern literature.

Astronomical consequences of the dark matter concept.

V. G. Fessenkov Astrophysical Institute, during 2012 year.

Proposing the simplest form of central symmetric potential celestial – mechanical type, i.e. $U(\varphi)=-\mu/\varphi$; clarifying that positions of images of lensing objects depend on profiles of dark matter and baryonic matter; searching the Cherenkov radiation in a galaxy's halo of dark matter, described by profiles of Navarro-Frenk-White and Burkert.

The authors argue that the standard field scalar potential couldn't be widely used for getting adequate curve lines of the galaxy and determining the profiles of dark matter in their halo. To discover the global properties of scalar fields that can describe the observable characteristics of dark matter on cosmological space and time scales, they proposed the simplest form of central symmetric potential celestial – mechanical type, i.e. $U(\varphi)=-\mu/\varphi$. It was shown that this potential allows the authors to attain dark matter profiles and rotational curve lines for dwarf galaxies. They also point out that these are in good agreement with some previous results, based on the N-body simulation method.

The light rays' propagation is searched in the spherical-symmetric gravitational field of the Galaxy described both by dark matter as well as baryonic matter. The authors focused on the case when contributions on the gravitational lensing of these components are comparable to each other. It is shown that positions of images of lensing objects depend on profiles of dark matter and baryonic matter. It was also argued that positions of images from the gravitational lens are drastically smaller than the distance between the gravitational lens and observer.

The Cherenkov radiation in a galaxy's halo of dark matter, described by profiles of Navarro-Frenk-White and Burkert was also searched. Powers of the Cherenkov radiation for the refractive indexes of gravitational fields with these profiles were calculated. It was shown that correspondent radiation temperatures in the X-rays diapason have the magnitude of about $T \sim 10^{-6} K$. It was also marked that its detection may be the criteria for choosing the preferable dark matter density distribution in a galaxy.

Some new astronomical consequences of the dark matter concept have been considered in detail. It was shown that the authors'observable predictions lay in the threshold of modern astronomical measurements.

Quantum theory was developed at a time when only "ordinary" matter was thought to exist. Its interaction with dark matter or dark energy is similarly absent in the many interpretations of the equations. Moreover, most of these interpretations postulate that reality in the quantum domain, especially causality, operates in a way that seems very strange in relation not only to common sense but also to the other sciences.

As described in Chapter 3, the interpretation that was dominant for the first half-century of quantum theory, the Copenhagen interpretation, holds that quantum reality is inherently probabilistic. In parallel to the dual nature of matter, wave-like or particle-like in different settings, a probability distribution that is derived from deterministic causation (Schrödinger's equation) "collapses" to give a single outcome. In addition, a central place is given to measurement in bringing this about. Quantum reality is thus held to be radically different from that in all other domains, including the remainder of science, in which the prevailing view is that nature and its causal relationships exist independently of the observer, i.e. they have their own prior reality. It is however possible to give an account of the subatomic domain that is realist in this sense, and that also reproduces all the predictions of quantum mechanics. This requires taking non-locality seriously, but there is in any case abundant evidence for this being an established feature of the quantum world. The existing theoretical tradition of this type, originating in the work of de Broglie and Bohm, does not command wide acceptance in physics, but there is scope for it to be developed in a manner that takes account of dark matter and/or energy.

Dark matter was originally identified at the macroscopic level from discrepancies in cosmological measurements and from gravitational lensing, reinforced by additional evidence, e.g. from the Bullet Cluster. At the microscopic level, various types of particle have been proposed as constituting dark matter, including supersymmetric neutralinos or gravitinos, the Kaluza-Klein particle, and axions. The necessary properties of such a particle, aside from gravitational effects, include that it does not interact with electromagnetic radiation such that its macro properties would be affected; there is no proscription of interaction with normal atomic matter. This raises the intriguing possibility that dark matter particles may participate in quantum phenomena. Far less is known about dark energy, and it is possible also that it could interact with normal matter so as to generate the phenomena of quantum physics.

In recent decades, causal analysis has focused on a broad range of scientific topics in disciplines other than physics, notably biology. The nature of causation is now much better understood. Briefly, a causal relationship involves both a mechanistic and a difference-making (e.g. mathematical) aspect, the two aspects being complementary. This perspective also illuminates apparently probabilistic behaviour: rather than being seen as resulting from an intrinsically different type of causation from deterministic behaviour, it reflects multiple/partial causation. In situations as different as coin-tossing and disease risk, the outcome of a conjunction of forces cannot be accurately predicted as it can in the deterministic situation. This is partly a question of level, the individual outcome being determined at the micro level but only visible at the macro level, where it is apparent as a particular probability at the level of a "population" of coin tosses or lung cancer cases. It is also a manifestation of how multiple causes combine to produce a particular outcome, e.g. the way that long-term heavy cigarette smoking causes lung cancer, but only in conjunction with many other causes, most of which are unknown.

If one assumes that the fundamental way the world works is the same in all domains, albeit with some important domain-specific differences, the findings of modern causal analysis can be used to illuminate the role of probabilistic causation in fundamental physics, recasting it as multiple causation. In particular, making the connection between dark matter/energy and quantum theory could help resolve some of the hitherto intractable problems in the latter, while providing new information on the former.

The use of superheated liquids in the search for astroparticle dark matter has recently begun to make a significant impact in the international dark matter search effort. Employed by only three projects worldwide (COUPP, PICASSO, SIMPLE), the advantage of the technique is that the thermodynamic operation of the devices requires radiations with linear energy transfer > 100 keV μm^{-1} to induce a signal, rendering the detector effectively insensitive to the majority of traditional detector backgrounds which plague more conventional dark matter search detectors (including photons, electrons and muons). This insensitivity is not trivial, translating to a reduced data record, and simplified, more efficient analyses of the results. Acoustic detection of the resulting gas bubble permits isolation of particle-induced events from a significantly reduced data sample, as well as nuclear recoil event discrimination. The simplicity of the detectors and instrumentation provides a relatively inexpensive and competitive search approach. Chapter 4 describes the salient aspects of both the technique and the three dark matter experiments which employ it, using SIMPLE as a point of reference.

Modern cosmology has many unresolved theoretical problems dealing with the very early, present and late time evolution of the Universe. One of these unresolved and tough problems is the cosmological constant and the recent cosmic accelerated expansion detected via observations of supernova type Ia and Cosmic Microwave Background radiation. It is not very clear to cosmologists what kind of energy-matter is driving this accelerated expansion, however numerous theoretical candidates have been proposed in the literature. Given the present unexplained state of the Universe, we face a deeper uncertainty in predicting the future cosmic evolution. Cosmologists are puzzled over an enigma why the energy densities of dark matter and vacuum energy are nearly the same at the present cosmic time, so-called the Cosmic Coincidence Problem (CCP). In Chapter 5, the authors provide a review of a phenomenological model which resolves CCP. This model involves a gravitational interaction between dark energy and dark matter in the background of a Friedmann-Robertson-Walker spacetime. The interaction ensures their energy densities are comparable at the present time irrespective of generic initial conditions of the Universe. Besides general relativity, the authors investigate the interacting dark energy model in alternative gravitational theories such as loop quantum gravity, generalized teleparallel T gravity and Horava-Lifshitz gravity. The authors briefly discuss thermodynamical aspects of such a physical process.

The study of the rotation curves of spiral galaxies reveals a nearly constant cored density distribution of Cold Dark Matter. N-body simulations however lead to a cuspy distribution on the galactic scale, with a central peak. A Bose-Einstein condensate (BEC) of light particles naturally solves this problem by predicting a repulsive force, obstructing the formation of the peak. After succinctly presenting the BEC model, in Chapter 6, the authors test it against rotation curve data for a set of 3 High Surface Brightness (HSB), 3 Low Surface Brightness (LSB) and 3 dwarf galaxies. The BEC model gives a similar fit to the Navarro-Frenk-White (NFW) dark matter model for all HSB and LSB galaxies in the sample. For dark matter dominated dwarf galaxies the addition of the BEC component improved more upon the purely baryonic fit than the NFW component. Thus despite the sharp cut-off of the halo density, the BEC dark matter candidate is consistent with the rotation curve data of all types of galaxies.

In Chapter 7, the authors review some developments of the Weyl tensor tool in the explanation of the galaxy flat rotation curves within the braneworld context. The authors also present a qualitative discussion about the sign of the so-called pressure parameter.

Newly-determined mass of $1.97 \pm 0.04 M_{\odot}$ for PSR J1614-2230 has been a challenge for the neutron star with a hyperon core (namely hyperon star), since hyperons usually reduce the theoretical maximum mass of the star. In Chapter 8, we consider dark matter as another possible constituent in hyperon stars' interior to lose this mass constrain. The authors take dark matter as self-interacting Fermi gas with certain repulsive interaction among the dark matter particles and assume non-interaction between dark matter and ordinary matter in the neutron star system. The authors find that the star maximum mass is sensitive to the particle mass of dark matter, and a high enough star mass larger than $2M_{\odot}$ could be achieved when the particle mass is small enough. In this particular model, a strong upper limit *0.64* GeV for dark matter mass is obtained in strongly-interacting dark matter and *0.16* GeV for dark matter mass in weakly-interacting dark matter. The latter is commonly considered it is the case for typical dark matter particles. Dark matter accumulated around the star could also contribute to the mass measurement, however, such contribution could be safely ignored when the generally used dilute dark matter density is assumed.

Instead of fluid type dark matter (DM) models, axion-like scalar fields with a periodic self-interaction are analyzed as a model of galaxy halos. Such a cold *Bose-Einstein* type condensate can provide a viable model for a *soliton type* interpretation of the DM `bullets' observed by means of gravitational lensing in merging galaxy clusters. Chapter 9 studies solitary waves for several self-interaction potentials in the relativistic Klein-Gordon equation, mainly in lower dimensions, and comment on their prospective shape-invariant collisions.

In: Research Development in Dark Matter Research ISBN: 978-1-62948-010-7
Editors: Nori Kinjo and Akira Nakajima

Chapter 1

SUBMICROSCOPIC VIEWPOINT ON GRAVITATION, COSMOLOGY, DARK ENERGY AND DARK MATTER AND THE FIRST DATA OF INERTON ASTRONOMY

Volodymyr Krasnoholovets *
Indra Scientific, Square du Solbosch, Brussels, Belgium

Abstract

The so-called Standard Model of particle physics and the Standard Model of cosmology have so far not been able to suggest unambiguously what the very nature of dark matter is. These Models are therefore flawed at the fundamental level. We propose a deeper fundamental approach, the submicroscopic deterministic concept that considers the real space as a substrate constituted of primary topological balls arranging a tessellation lattice. Dark energy is accounted for by a peculiar distribution of topological balls inside the universe considered as a huge-range cluster in the total tessel-lattice. A moving particle interacts with the space generating spatial excitations named inertons by the author. Inertons being a substructure of the matter waves are present in any material object owing to the oscillating motion of its entities. Inertons form a vast series of harmonics that in the form of standing spherical waves spread far

* E-mail address: v_kras@yahoo.com

beyond the object inducing its Newton's gravitational potential. Starting from first submicroscopic principles the submicroscopic concept explained all phenomena predicted by general relativity. Dark matter is derived as a viscosity of space induced by standing inerton waves in the surrounding of star clusters. Two devices that measure inerton signals, which were designed by our team, are described. Some measurements of the Earth and Sun inerton signals conducted by using these devices are demonstrated. Finally we suggest designing the first inerton measuring complex, an inerton observatory, which will be able to measure amplitude, spectral and time characteristics of non-stationary inerton signals received from outer space.

1. Introduction

There are a number of review articles dedicated to the problem of dark matter (e.g. Garret and Dūda 2011). But the problem still looks as difficult for conventional cosmology and fundamental physics, as the notion of "dark matter" is not organically part ofany of the main concepts. Researchers try to relate dark matter to massive objects in the universe rather artificially, because this enigmatic substance cannot be derived directly from the main equations that currently describe the fundamentals of gravitational physics and cosmology. That is why, in order to understand and solve the problem of dark matter, we have to run through all three levels: descriptive, analytical and conceptual.

The descriptive level means astronomical observations and these are abundant: dark matter indeed was observed (Gilmore et al. 2007; Einasto 2010; Toffolatti et al. 2012; Benítez-Llambay et al. 2013).

The analytical level is represented by many diverse views, theories and experiments. Among major hypotheses we can name: MOND, which has actively been supported by a number of astronomers (Famaey and McGauph, 2012; 2013); cold dark matter (CDM) with the further broadening to the Lambda-CDM model, which includes CDM and parameterizes the Big Bang cosmological model having a cosmological constant Λ (Tinker et al. 2008; Boylan-Kolchin 2011) with its hypothetical weakly interacting particles – WIMPs (Merritt and Bertone 2005), warm dark matter (Stasielak 2006; de Vega and Sanchez 2011) whose particles would be sterile neutrinos (Boyarsky et al. 2009; Abazajian et al. 2012), which complicates the situation, and hot dark matter with neutrinos as carriers.

In particular, it is emphasized (Kamada et al. 2013)that the validity of the ΛCDM model on the galactic and the subgalactic scales has long been caught up in debate and that some researches argue that the number of dark matter subhalos is 10−100 times larger than the number of satellites observed around the Milky Way; so they argue that missing massive satellite problem can be solved only in the framework of warm dark matter in which carriers of the missing mass are sterile neutrinos with an energy about 2 keV. Quite recently de Vega et al. (2013) have considered the possibility of searching for sterile neutrino signatures in two experiments that are running at present, MARE and KATRIN, focused on the Rhenium and Tritium beta decays respectively. Hence they anticipate that sterile neutrinos will interact with a nucleus (or some nuclei) in the detector.

On the other hand, other researchers (Zolotov et al. 2012; Brooks et al. 2013) in support of the ΛCDM model and using super computers could show that the inclusion of stars that go supernova can alter the dark matter structure of galaxies driving dark matter out of the centers of satellites and this is the mechanism of lowering of the masses of the bright satellite galaxies; this brings the predicted and observed masses of the Milky Way's satellites into agreement.

The motivation for searching for WIMP is that cosmological observations require dark matter that is composed of a new kind of unknown particles. The weak force for their interaction requires new states as well. It is believed that WIMPs are naturally produced as thermal relics of the Big Bang with the densities required for dark matter; this WIMP miracle drives most dark matter searches (Feng and Kumar 2008). Referring to Ya. Zeldovich and his followers Feng and Kumar (2008), they point out that a particle's thermal relic density is

$$\Omega_X \propto 1/\langle\sigma\upsilon\rangle \sim m_X^2 / g_X^4,$$

where $\langle\sigma\upsilon\rangle$ is its thermally-averaged annihilation cross section, m_X and g_X are the characteristic mass scale and coupling entering this cross section, and the last step follows from dimensional analysis; m_X is associated with the dark matter particle's mass; the relic density is typically within an order of magnitude of the observed value, $\Omega_X \approx 0.24$. Their further analysis discloses a wide possible range for the mass: 10 MeV $\leq m_X \leq$10 TeV.

Since WIMPs do not experience electromagnetic interactions, they have to pass right through solid matter, for example a detector. The cross-section of their interaction with an atomic nucleus is considered close to that of the weak force, less than 10^{-40} cm^2. This means that WIMPs may very rarely hit an atomic nucleus very rare; nevertheless, such signature could be measured experimentally. Rocha et al. (2013) note that SIDM models with the cross-section per mass ratio ~ 0.1 cm^2/g consistent with all observational constraints.

However, so far no real success has been reached in the detection of WIMPs, which are so important for the ΛCDM model. In most cases the experiments searching for these hypothetical dark matter particles have become unsuccessful, though some others seem promising for further research. Some of these experiments are: the XENON dark matter search (Aprile et al. 2011), the Cryogenic Dark Matter Search (CDMS) detector (CDMS Collaboration, 2011), the DAMA/LIBRA detector (Bernabei et al. 2010; 2012),and others.

McCulloch (2012) suggested a model that modifies inertial mass, such that it correlates the gravitational interaction between objects correctly describing the Tully-Fisher relation, i.e. the anomalous behavior of the velocity of stars different from Kepler's law. Some researchers try to introduce new elements into the description of gravitating objects, such that dark matter is excluded automatically. For example, the physical vacuum might be considered as a fluid of virtual gravitational dipoles and the phenomenon of dark matter might be explained by the gravitational polarization of the quantum vacuum by the known baryonic matter (Hajdukovic 2011). On the other hand, by another hypothesis, strict conformal symmetry (local Weyl scaling covariance $g_{\mu\nu}(x) \to g_{\mu\nu}(x)\,\Omega^2(x)$, for a scalar field $\Phi(x) \to \Phi(x)\Omega^{-1}(x)$), might be postulated for all elementary massless fields, retaining standard fermion and gauge boson theory though with a modification of Einstein–Hilbert general relativity, the Friedmann equations and the Higgs scalar field model, with no new physical fields (Nesbet 2013). Conformal gravity and a conformal Higgs model remove trace elements, which are the major elements in the general relativity formalism, and such an approach allows one to fit empirical data on galactic rotational velocities, galactic halos, and Hubble expansion including dark energy without invoking dark matter.

The conceptual level just partly started by those last mentioned works.

Understanding the structure and nature of space are highly important, because they must be the foundation for the basic theoretical concepts of modern physics. Dark energy and dark matter introduce a dissonance in modern physics, which means that something is wrong with the fundamentals.

Dark energy is a constant energy density uniformly filling the space of the universe (i.e., postulated as a non-zero vacuum energy and pressure). The essence of dark energy is a subject of dispute. It reacts insignificantly with ordinary matter with the exception of gravity, has low density and a very homogeneous distribution.

In the USA the Dark Energy Task Force (DETF) was established by the Astronomy and Astrophysics Advisory Committee (AAAC) on future dark energy research. The DETF notes: "Dark energy appears to be the dominant component of the physical Universe, yet there is no persuasive theoretical explanation for its existence or magnitude. The acceleration of the Universe is, along with dark matter, the observed phenomenon that most directly demonstrates that our theories of fundamental particles and gravity are either incorrect or incomplete.

Most experts believe that nothing short of a revolution in our understanding of fundamental physics will be required to achieve a full understanding of the cosmic acceleration. For these reasons, the nature of dark energy ranks among the very most compelling of all outstanding problems in physical science. These circumstances demand an ambitious observational program to determine the dark energy properties as well as possible" (Albrecht et al. 2006).

Kroupa (2012) writes about the dark matter crisis and falsification of the current standard model of cosmology; namely, he notes that the number and dark matter halo mass distribution of the Milky Way's satellite galaxies is in highly significant disagreement with the expectations from the standard model of cosmology. There is no physically known process that may be able to solve the disagreements. He cites a long list of 22 complaints about the standard model of cosmology. Then he notes that the need to introduce dynamically relevant dark matter on galactic scales is required because the assumption that Einstein's field equations of 1916 be valid on galactic and cosmological scales led to a failure of them as soon as kinematical measurements in galaxies and of galaxies in galaxy clusters became available long after 1916. Kroupa (2012) points out that the speculation that exotic dark matter particles exist cannot be understood within the standard model of particle physics, as they have not been discovered by direct experiment despite a highly significant effort all around the world. Besides, he emphasizes that an effective gravitation appears to be non-Einsteinian and non-Newtonian, in particular, the true solution to gravitation may be Milgrom's suggestion that the effective gravitational force law at ultra-low accelerations $a_0 \approx 1.2 \times 10^{-10}$ m·s^{-1} (3.6 pc/Myr^2) (Milgrom 1983; 2009). No dark matter but rather modified gravity,modified Newtonian dynamics (MOND).

Incorporation of a MOND acceleration constant, which varies with redshift and an expansion history, and the introduction of neutrinos particles, allows one to describe experimentally observed galaxies that cannot be written by the usual Friedmann models (Angusand Diaferio 2011). Thus, no needs for peculiar dark matter particles behind the standard model of particle physics.

Kroupa et al. (2012) continuous to criticize the standard model of cosmology, which as they underline, is based on five postulates: 1) cosmological physics is based on assuming Einstein's field equation to hold on galactic and larger scales and for very small accelerations as are found in galactic systems; 2) all matter emerged in the Big Bang. The authors signify that these two postulates lead to an inhomogeneous and highly curved cosmological model, though the observed distribution of matter is rather homogeneous and flat; 3) an inflation is additionally postulated to drive the universe to near flatness and homogeneity briefly after the Big Bang; 4) the structures and their kinematics based on this postulate does not work correctly until cold or warm dark matter is hypothesized; 5) in addition, such model still in addition requires dark energy to driving the inflation.

Kroupa et al. (2012) conclude that cosmology requires a new paradigm. They focus on the extraordinary success of MOND in accounting for the observational data on galactic scales and its properties suggest that the expected underlying theory will contain a deep connection between the dynamics within local systems and the state of the universe at large.

Cosmic microwave background radiation discovered in 1964 (Penzias and Wilson 1964) is considered as the most important argument in support of the Big Bang model of the Universe. Since that the hot Big Bang model gradually gained popularity and pushed aside both the early Steady State models (Hoyle et al. 1993; 1994) and the ether theories (e.g., Thronhill 2001).

If the cosmic microwave background has the spectrum of a blackbody, then the universe might be treated as originated from a hotter, denser phase. Results on cosmic microwave background are considered to be of paramount important for cosmology. The NASA's COBE (COBE 2009) and WMAP satellites (WMAP 2011) and the Planck satellite of the European Space Agency (Rocha 2010) showed that the cosmic microwave background spectrum is that of a nearly perfect blackbody with a temperature of 2.725 K. The researchers claim this observation matches the predictions of the hot Big Bang theory.

However, these satellite results were strongly criticized (Robitaille 2007; 2009; 2010a; 2010b). Robitaille clearly points out that those experiments are fraud and the satellites' infrared antennas recorded signals of the Earth's oceans. He writes (Robitaille 2007): "the obscured resonances at ~4.63 ppm in the water spectrum would still have a signal to noise of ~5:1, if the water line had not contaminated this region... For WMAP, the signal to noise is less than 2:1, and the signal of interest is located at the same frequency of the contamination… Signal suppression, by a factor of 100, or more, while still viewing the underlying signal, depends on the ability to control the source. This has been verified in numerous laboratories where the sample is known and where the correct answer can be readily ascertained. As such, it is impossible for the WMAP team to remove the galactic foreground given the dynamic range situation between the contaminant and the signal of interest… Relative to signal to noise, the WMAP team is unable to confirm that the anisotropic "signal" observed at any given point is not noise… The requirements for image stability in cosmology are well beyond the reach of both COBE and WMAP." "…with the aid of the Planck satellite, the electromagnetics laboratories of the world should be able to confirm or refute the existence of a ~3 K cosmic signal. The key to this puzzle rests in the understanding of the LFI (low frequency instrument) and reference targets...

Enough evidence is already beginning to build … indicating that physics, astrophysics, and geophysics stand on the verge of a significant reformulation. In any event, the definitive proof that the monopole of microwave background belongs to the Earth has now been provided." (Robitaille, 2010b).

I often asked experimentalists at the conferences dealing with gravitational physics and cosmology: "Why do you think that the recorded signal, which you have discussed here, came from the sky but not from a neighboring room?" I have never heard an answer.

The criticism (Robitaille 2007; 2009; 2010a; 2010b) is quite serious and means that the existence of the cosmic microwave background radiation is very debatable. In any case such a radiation, if any, does not have a spectrum typical for the spectrum of a blackbody, which automatically eliminates the scenario of a hot beginning of the universe.

Are there any new ideas for the developing of cosmology? An interesting study carried out by Volovik (2003) in which he hypothesizes regarding the universe as a Helium droplet. Volovik's vacuum is similar to the ground state of quantum liquids; the "relevant ground state energy" that gravitates is $E_0 - \mu N$ not E_0; the universal equation of state for the vacuum of the universe (the same as for the quantum liquid): $\rho_{\text{vac}} = -P_{\text{vac}}$ and this equation exactly coincides with the equation of state of the cosmological constant in general relativity. This is mathematically equivalent to a zero-energy superfluid, although its fine morphology has gone unrecognized.

The constancy of the speed of light is the pillar of special relativity. It is believed the constancy of the speed of light takes place in the four dimensional space-time whose space-time dimension number (four) is constant. However, switching of a possibility for the slicing of extra space dimensions, which provided by varying dimension numbers, gives rise to the emergence of a fine space structure. Varying dimension numbers are derived from varying speed of light theories (Barrow 2003; Magueijo 2003; Casado 2003; Ellis and Uzan 2003; Petit and d'Agostini 2007); the speed of light is treated as a parameter that varies in time. There are arguments demonstrating that the value of c is not a constant in vacuum (Santilli 2006). On the other hand, the possible variation of c in the context of the present accelerating universe is not permitted under the variable Λ models (Ghosh et al. 2012).

So we can see that dark matter, Big Bang, dark energy and cosmology as a whole are veiled with an impenetrable turbidity. On the other hand, a theory of dark matter recently proposed by the author (Krasnoholovets 2011) was based on a detailed mathematical theory of the ordinary physical space and the concept of subquantum physics developed in the strictly defined space with its peculiar structure, rules and properties. Below, referring to the physical meaning of cosmology, we consider the Einstein equations, features of metrics and how matter and the gravitational laws emerge in the space considered as a tessellated lattice of primary topological balls (section 2), then consider the Friedmann solutions and show how dark energy emerges in the universe constituted as the tessel-lattice (section 3), after that we demonstrate how in the tessel-lattice gravitational laws resolve the dark matter problem (section 4). Then we exhibit our devices that measure flows of mass and provide the reader with our first experimental data of astrophysical measurements of massive signals (section 5).

2. Einstein Field Equations, Metrics and the Gravitation

2.1. General Relativity

In the general theory of relativity the Einstein-Hilbert equations, or more exactly a set of 10 coupled nonlinear partial differential equations,

$$R_{\mu\nu} - \frac{R}{2} g_{\mu\nu} + \Lambda g_{\mu\nu} = \frac{8\pi G}{c^4} T_{\mu\nu} \tag{1}$$

connect the metric tensor $g_{\mu\nu}$ of the curled space-time with properties of matter that fills the space-time and is characterized by the stress-energy tensor $T_{\mu\nu}$. Here, $R_{\mu\nu}$ is the Ricci curvature tensor, R is the scalar curvature, Λ is the cosmological constant, G is Newton's gravitational constant, c the speed of light in vacuum. Components of the metric tensor $g_{\mu\nu}$ represent the gravitational potential in the chosen point of space-time.

Since both the Ricci tensor $R_{\mu\nu}$ and scalar curvature R depend on the metric in an intricate nonlinear way, Eq. (1) are really equations for the metric tensor $g_{\mu\nu}$. That is why researchers first of all try to investigate the metric tensor.

The classical example is the Schwarzschild metric (Schwarzschild 1916). The idea is to consider the square of a linear element of space-time ds^2 that does not change its form when turning spatial coordinates ξ^i around the axis that passes through the origin of coordinates. In the flat Minkowski space-time the square of the linear element in the spherical coordinates

$$ds^2 = -c^2dt^2 + d\rho^2 + \rho^2 d\phi^2 + \rho^2 \sin^2\phi\, d\theta^2, \tag{2}$$

which is generalized in general relativity by using the expression

$$ds^2 = -A(r)c^2dt^2 + B(r)\,dr^2 + r^2 d\phi^2 + r^2 \sin^2\phi\, d\theta^2. \tag{3}$$

The functions $A(r)$ and $B(r)$ include the gravitational field of the central mass M as the Newtonian potential $-GM/r$, which after a long chain of transformations and calculations of the Christoffel symbols $\Gamma^{\beta}_{\mu\nu}$ and the Ricci tensor $R_{\mu\nu}$ becomes finally (in the presentation of the Hilbert's metric)

$$ds^2 = -\left(1 - \frac{2GM}{c^2 r}\right)dt^2 + \left(1 - \frac{2GM}{c^2 r}\right)^{-1} dr^2 + r^2 d\phi^2 + r^2 \sin^2\phi\, d\theta^2. \tag{4}$$

The majority of modern researchers associate a coordinate singularity in the Schwarzschild metric (4) with the existence of so-called black holes in the universe, which ostensibly have been observed by astronomers. On the other hand, some other investigators of general relativity strongly oppose the existence of black holes showing that black holes are not a consequence of the theory of general relativity (Crothers 2005, 2010, 2011, 2012; Loinger 2002, 2007; Rabounski 2008).

The Schwarzschild metric is not unique. There exist a number of different metrics that satisfy the conditions of invariance of the square of a linear element of space-time ds^2 at transformation of coordinates that does not change the form of the element. For example, one of them is the Nordström metric $ds^2 = e^{2\varphi}\eta_{\mu\nu}dx^\mu dx^\nu$ where $\varphi = -GM/(c^2 r)$ is the gravitational potential and $d\sigma^2 = \eta_{\mu\nu}dx^\mu dx^\nu$ is the line element for the Minkowski flat space-time. This metric is interesting, as it starts directly from Minkowski's. The exact solution looks as follows

$$ds^2 = \left[1 - GM/(c^2\rho)\right]\left[-dt^2 + d\rho^2 + \rho^2(d\theta^2 + \sin^2\theta\, d\phi^2)\right]. \tag{5}$$

By using the transformation $r = \rho \cdot [1 - GM/(c^2\rho)]$, the metric changes into

$$ds^2 = [1 + GM/(c^2 r)]^{-2}(-dt^2 + dr^2) + r^2(d\theta^2 + \sin^2 O\, d\phi^2). \tag{6}$$

Considering a Nordström type metric, a Schwarzschild like metric, which uses exponential functions, can be presented in the form (Baretti Machín 2012)

$$ds^2 = \exp[-2GM/(c^2 r)]c^2 dt^2 - \exp[2GM/(c^2 r)]dr^2 - r^2 d\theta^2 - r^2 \sin^2\theta\, d\phi^2 . \quad (7)$$

The metric (7) is specified with the absence of a coordinate singularity and at the same time expanding exponential factors, since all planetary motion takes place in the region where $2GM/(c^2 r) << 1$, we arrive at the Schwarzschild metric (4).

The Schwarzschild-Hilbert metric has advantages over others (Kruskal, Szekeres, Gödel and so on), because it can correctly describe classic experiments, such as the movement of the perihelion of Mercury, the bending of a light ray by the sun, the red shift spectral lines in the presence of the gravitational field, and the radio signal delay (the Shapiro effect).

Nevertheless, owing to the obvious success of general relativity, the problem of understanding gravitation still remains. Once again, eqs. (1) describe a connecetion between the matter and the space curved by this matter, though the gravitational potentials are concentrated in the metric components.

The most fundamental problem is how general relativity can be reconciled with the laws of quantum physics to produce a complete and self-consistent theory of quantum gravity. It seems this challenge cannot be resolved in principle because of principal differences in approaches to physical laws by microscopic quantum physics and phenomenological general relativity. Besides, general relativity does not look like a true physical theory but rather like an abstract mathematical theory.

In the case of general relativity, which abandoned the classical ether concept and introduced an abstract vague vacuum, we can distinguish six problems, *conceptual difficulties*, which do not have resolutions in the framework of relativity formalism:

1. a massive object can influence space-time but cannot be derived from it, because the unknown and undetermined parameter mass$M$ is entirely separated from the phenomenological notion of space-time;
2. the equation (1), which connects space-time and matter, includes a metric that should be chosen by hands, i.e. artificially (for example, expression (3));
3. the formalism of relativity is failing on a microscopic scale as it does not pay attention to the wave nature of matter; for instance, one can see this

in a recent work(Salih et al. 2012) in which the authors simply decrease the scale to microspace but do not touch the wave nature of the microscopic world. At distances equal compared to or less than the object's de Broglie wavelength the formalism of general relativity has to give way to an approach based on microscopic consideration;

4. general relativity is founded on the basic Newtonian term $-GM/r$, but cannot explain its origin;
5. general relativity does not offer any sort of particles/quasi-particles, which will be able to realize short-range action in the gravitational attraction of objects and hence it is a theory based on an action-at-a-distance phenomenological approach with the speed c, the same as the Newtonian theory (this one acts immediately, and also quantum mechanics whose long-range action also falls within the range of its conceptual difficulties); regarding the quasi-particles *gravitons* we can say that, based on the studies of other researchers (Loinger 2007) as well as experimental results (Krasnoholovets and Byckov 2000), these are abstract mathematical objects absent in real nature;
6. light, which plays an exceptionally important role in relativity, has to be massless in the theory; however, light carriers, photons, transfer momentum and energy and therefore by the principle of equivalence, photons must have non-zero mass;
7. it is complete unclear how to involve the formalism of general relativity in the study of gravitational interaction of an ensemble of massive points when the pair potentials must be paramount impotent.

In addition to these critical items we should mention that the formalism of general relativity does not include dark matter and dark energy in its consideration.

2.2. Submicroscopic Concept of Gravity

2.2.1. Space Structure, Generation of Matter and the Quantum Laws

The submicroscopic concept based on the constitution of physical space (Bounias and Krasnoholovets 2003; 2004) and submicroscopic mechanics (Krasnoholovets and Ivanovsky 1993; Krasnoholovets 1997, 2002), which allow a detailed theory of gravity to be derived introducing a radically new approach to the problem of gravitation and quantum gravity in particular.

The theory of physical space was built as a pure mathematical theory based on set theory, topology and fractal geometry (Bounias and Krasnoholovets 2003; 2004). The theory shows the necessity of the existence of a primary topological ball. Balls completely fill space forming a tessellation lattice named the tessel-lattice. The tessel-lattice is provided with special rules. The size of a cell equals the Planck size $l_{\text{Planck}} = \sqrt{\hbar G / c^3} \approx 1.162 \times 10^{-35}$ m.

Classically, a measure is a comparison of the measured object with some unit taken as a standard. This signifies that one can introduce matter through a change in a cell of the tessel-lattice. Hence a local deformation of the degenerate tessel-lattice, which emerges at switching a fractal volumetric deformation of a cell, can be associated with the physical notion of mass. Such deformation can migrate from cell to cell or can scatter by cells. A stable symmetric volumetric deformation of a cell is associated with the creation of a massive particle. Thus mass is identified with a change of the volume of a fractal-deformed cell: The mass m_A of a particulate ball A is a function of the fractal-related decrease of the volume of the ball:

$$m_A \propto (V^{\text{deg. cell}} / V^{\text{part}} - 1) \cdot (\mathrm{e}_{\text{fract}} - 1)_{\mathrm{e}_{\text{fract}} > 1} \tag{8}$$

where $V^{\text{deg. cell}}$ is the typical average volume of a cell in the tessel-lattice in the degenerate state; V^{part} is the volume of the kernel cell of the particle; (e) is the Bouligand exponent, and $(\mathrm{e}_{\text{fract}} - 1)$ the gain in dimensionality given by the fractal iteration. Just a volume decrease is not sufficient for providing a ball with mass, since a dimensional increase is a necessary condition (there should be a change in volumetric fractality of the ball).

Note that other researchers also point to a discrete space: "this is the effervescent Grid… Matter is not what it used to be. It consists of small, more-or-less stable patterns of disturbance in the Grid…Usually the metric field is taken to be fundamental, but in many ways it resembles a condensate, and that view of it may become important…What we ordinarily call matter consists of more-or-less stable patterns of excitation in the Grid, which is more fundamental. At least, that's how things look today" (Wilczek 2009).

So, the physical space is treated as a real substrate with its inner discrete structure, the tessel-lattice. The degenerate state of the tessel-lattice can be correlated with the physical notion emptiness. The tessel-lattice becomes the origin of matter; therefore, we may start from *the theorem of something* and determine major concepts fundamental physics operates with, instead of the usual *the theorem of everything* that tries to unify all fundamental interactions without understanding the basic figurants of these interactions.

The interaction of a moving particle-like deformation with the surrounding lattice involves a fractal decomposition process.

Submicroscopic mechanics of a particle moving in the tessel-lattice was developed in a series of works (Krasnoholovets and Ivanovsky 1993; Krasnoholovets 1997, 2002). Poincaré (1906) and de Broglie (1925) hinted to the principles of the motion, but the readers did not take their words into account. In particular, Poincaré noted that a particle, as a singularity of the ether, moves in it surrounded by the ether's excitations. De Broglie suggested that a point particle is guided by a real phase wave and he introduced two major relationships for a particle accompanied with such wave

$$E = h\nu, \qquad p/h = \lambda, \tag{9}$$

where ν and λ are the wave's frequency and wavelength, respectively, and E and p are the particle's energy and momentum. However, other physicists called such union of two different physical entities a 'wave-particle'.

In submicroscopic mechanics a particle moving in the tessel-lattice interacts with its cells, which results in the generation of excitations named *inertons* (these excitations represent the particle's inertia). The Lagrangian of the submicroscopic mechanism includes three major parts: the kinetic energy of the particle, the kinetic energy of the ensemble of inertons and the energy of interaction of the particle and inertons. Schematically the Lagrangian can be shown in the form

$$L = \tfrac{1}{2} m\dot{x}^2 + \tfrac{1}{2}\mu\dot{\chi}^2 - \frac{\sqrt{m\mu}\,\dot{x}\chi}{T} \tag{10}$$

where m and x are the particle's mass and the position, respectively, and μ and χ are the mass and the position of the inerton cloud of the particle, respectively; T is the period of collisions between the particle and its inertons.

The Euler-Lagrange equations based on the Lagrangian (10) indicate periodicity in the behavior of the particle. The solution for the particle exhibits oscillates the particle velocity between the initial value υ and zero along each section λ of the particle path

$$\dot{x}(t) = \upsilon \cdot \left[1 - |\sin(\pi t/T)|\right], \tag{11}$$

$$x = \upsilon t + \frac{\lambda}{\pi}\left\{(-1)^{[t/T]}\cos(\pi t/T) - (1 + 2[t/T])\right\}. \tag{12}$$

The particle's spatial amplitude is $\lambda = \upsilon T$. The same for the cloud of inertons: $\Lambda = cT$, where c is the speed of light – the velocity with which inertons are emitted from the particle at its collisions with oncoming cells of the tessel-lattice.

The solution for the inerton cloud of the particle shows that the cloud oscillates around the particle (Krasnoholovets 1997):

$$\dot{\chi} = (-1)^{[t/T]} c\cos(\pi t/T), \tag{13}$$

$$\chi = (\Lambda/\pi)\,|\sin(\pi t/T)| \tag{14}$$

The inerton cloud is emitted from the particle each odd section $\lambda/2$ of the particle path and during the next even section $\lambda/2$ these inertons return to the particle.

Amplitudes λ and Λ become connected by means of relationship

$$\Lambda = \lambda c/\upsilon. \tag{15}$$

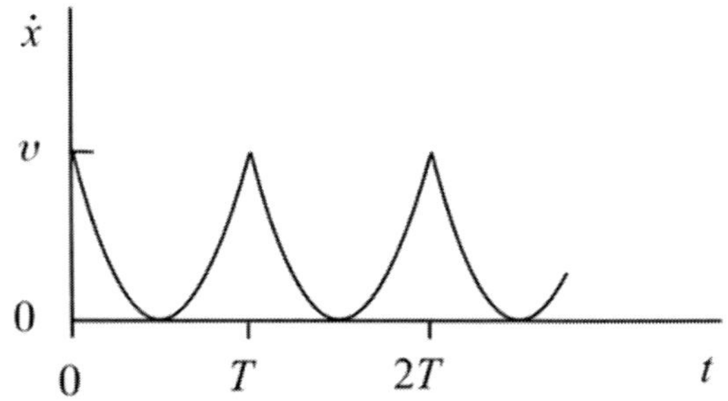

Figure 1. Velocity of a particle moving in the tessel-lattice versus time.

The amplitude of spatial oscillations of the particle λ appears in quantum mechanics as the de Broglie wavelength. The amplitude of the particle's inerton cloud Λ becomes implicitly apparent through the availability of the wave ψ-function. Therefore, the physical meaning of the ψ-function becomes completely clear – it describes the range of space around the particle, which is perturbed by the particle's inertons. The cloud's inertons play the role of carriers of the field of inertia of the particle.

The connection of such mechanics developed in the real space to quantum mechanics constructed in an phase abstract space becomes clear when the same problem is treated in the Hamilton-Jacobi presentation. The corresponding expression for the shortened action S_1 of the particle appears in the form

$$\frac{(\partial S_1/\partial x)^2}{2m} + \tfrac{1}{2}m\left(\frac{2\pi}{2T}\right)^2 x^2 = E \qquad (16)$$

which is the shortened action of a typical harmonic oscillator with a cyclic time period equal to $2T$. Introduction of the action-angle variables leads to the increment of the particle action within the cyclic period $2T$

$$\Delta S_1 = \oint p dx = E \cdot 2T. \qquad (17)$$

This expression can also be rewritten via the frequency $\nu = 1/(2T)$. Besides, $1/T$ is the frequency of collisions of the particle with its inerton cloud. Then using the relation $E = m\upsilon^2/2$ we get

$$\Delta S_1 = m\upsilon \cdot \upsilon T = p \cdot \lambda. \qquad (18)$$

If we equal the increment of action ΔS_1 to the Planck's constant h, we immediately obtain from expressions (17) and (18) the de Broglie's relationships (9).

This means that h is the minimal action of the periodic motion of a particle when the particle is guided solely by the tessel-lattice and any influence on the side of external fields or direct contacts with other physical systems does not disturb the particle.

The availability of relationships (9) allows one to obtain the usual Schrödinger equation (de Broglie 1986)

$$-\frac{\hbar^2}{2m}\nabla^2\psi(r, t) + V(r)\psi(r, t) = E\psi(r, t) \qquad (19)$$

In conventional quantum mechanics an undetermined 'wave-particle' is further substituted by a package of superimposed monochromatic abstract waves. It is this approximation that gives rise to the inequality $\Delta k\,\Delta x > 1$ for changes of the position Δx and the wave number Δk of the wave package under consideration; multiplying this inequality by the Planck's constant h, we derive the Heisenberg's uncertainty $\Delta p \Delta x > h$ (see p. 52 in de Broglie 1986). However, as we have just shown, such an interpretation in terms of 'wave-particle' is a very crude approximation to the real process of the motion, which realistically includes two subsystems – the particle itself and the inerton cloud that oscillates around the particle. This means that the origin of the quantum mechanical commutator

$$\hat{p}\,\hat{x} - \hat{x}\,\hat{p} = -i\hbar \tag{20}$$

is not in the uncertainties between Δk and Δx, but in the relationships (9) whose origin lies in the oscillating motion of the particle under consideration and its cloud of inertons in each section λ of the particle's path, which is characterized by the quantum of action $\hbar$.

Just to show that the uncertainty principle is not so fundamental, the reader may read the study of Nobel Prize winner Hans Georg Dehmelt (Van Dyck et al. 1976, 1986); in collaboration with other researchers he could, with high precision measurements, gauge a number of parameters of an electron practically at rest. Besides, in recent high-resolution scanning tunneling microscopy experiments it has been revealed that the interpretation of the density of electron charge as a statistical quantity leads to a conflict with the Heisenberg uncertainty principle (Hofer 2012); the uncertainty principle was violated by close to two orders of magnitude.

Submicroscopic mechanics was verified experimentally in many different physical systems. In was demonstrated that the Earth rotating from the west to east around its axis generates a flow of inertons, which we amplified in a simple resonator (Krasnoholovets and Byckov, 2000). Those amplified inerton waves influenced samples cut of razor blades, which were put in the resonator. The samples were studied by using an electron microscope. It was observed that the fine morphological structure of the edge of each razor blade indeed changed; a crude morphological structure remained the same. Hence the expected changes in the structure of the test specimens caused by the Earth's inerton field were in fact convincingly fixed in micrographs.

The submicroscopic concept started from the idea that electrons in atoms or in a metal should be treated as extended objects, but not point-like: an electron together with its inerton cloud has the length equal to their de Broglie's wavelength and the electron's inerton 'wings' spread up to the distance $\Lambda = \lambda c / \upsilon$ (15) in transversal directions around the particle, which for electrons in an atom and a metal is around 10 nm. Hence, the cross section of the electron's inerton cloud is $\sim \Lambda^2 \approx 100$ nm^2. Thus, such an object is able to absorb N photons simultaneously, which can be considered an anomalous photoelectric effect (Krasnoholovets, 2001a). The corresponding probability was calculated and was applied to describe tens of different experiments on generation of photoelectrons in gases and a metal. The results are completely satisfactory. Indeed, if the intensity of a laser pulse is 10^{16} to 10^{18} W/cm^2, we can estimate a mean distance between photons in the flux of the laser pulse as 3-4 nm. Then the number of photons, which bombard the electron's inerton cloud is $\Lambda^2 / d^2 \approx 10$. In other words, the size of the electron (jointly with its inerton cloud) is large enough and can absorb up to 10 photons from a laser flux simultaneously. The total energy of these 10 photons exceeds the ionized potential of atoms in a gas (or the work function in a metal).

The freezing of fluctuating molecules in aqueous solutions affected by an inerton field was demonstrated experimentally (Krasnoholovets, Skliarenko and Strokach 2006; Andreev et al. 2007).

Binding of electrons into electron droplets became possible only due to an elastic attraction of electrons in the electron droplet (Krasnoholovets, Kukhtarev, Kukhtareva, 2006). Our rigorous experimental results manifested that the formation of stable electron clusters containing about 10^{10}electrons occurred due to the absorption of inerton radiation by photoelectrons where the inerton radiation originated from the crystal surface illuminated by a laser beam.

2.2.1. Generation of Gravity

Submicroscopic mechanics considered above looks like a kinetic theory of a particle that collides with its inerton cloud.

The understanding of gravity starts from understanding the reasons why inertons emitted by a moving particle come back to it.

This is possible if inertons are not solid, but flexible quasi-particles. The same properties must be inherent in the case of the tessel-lattice. The interaction of a moving particle-like deformation with the surrounding lattice involves a fractal decomposition process. An inerton is generated at each collision of the particle with oncoming cells of the tessel-lattice: the particle loses a fragment of its mass, which is transferred into an emitted inerton. A number of inertons emitted by the particle passing the section $\lambda/2$ until the stop, is of the order of $\lambda/l_{\text{Planck}} \sim 10^{25}$ (if the particle's de Broglie wavelength $\lambda \sim 10^{-10}$ m). So inertons start having the velocity of light c and a mass m_{inert} that may vary from inerton to inerton (Krasnoholovets 2001b). At the distance of Λ from the particle an inerton stops having no mass but gaining a tension, which is demonstrated in Figure 2. This is a typical oscillating process when a deformation changes to a tension and vice versa. Then at the distance Λ the tessel-lattice as a whole pushes the inerton back to the particle and the inerton hopping from cell to cell acquires its mass again and finally is absorbed by the particle transferring the mass m_{inert} on to it.

Note that the situation with the particle's inertons is similar to the situation with the particle itself. Within the section λ, due to the emission and re-absorption of the particle's inerton cloud, all parameters of the particle undergo periodical changes (Krasnoholovets 2010): velocity $\upsilon \to 0 \to \upsilon$; mass $m \to 0 \to m$ and the tension $0 \to \Xi \to 0$; electric charge $e \to 0 \to e$ and the magnetic charge, i.e. monopole state $0 \to g \to 0$; particle shape: beanlike $\to$ spherical $\to$ beanlike (such internal motion manifests itself in conventional quantum mechanics as a half-integer spin, i.e. snin-1/2).

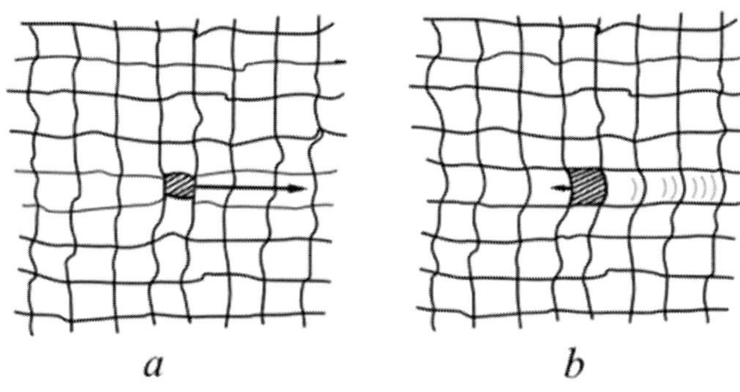

Figure 2. Inerton from the particle's inerton cloud: *a* – the inerton in the vicinity of the particle where it possesses a mass m_{inert} (the deformed cell); *b* – the inerton at the largest distance from the particle. Here, the inerton acquires a tension, or rugosity ξ_{inert} (the tense cell). This tension strains the tessel-lattice that due to the elasticity reflects the inerton back to the particle.

So since inertons transfer fragments of the particle's mass, they become not only carries of inert properties of the particle but also play the role of carriers of gravitational properties of the particle. First of all we should describe how inertons irradiated by the particle come back to it returning fragments of its mass as well as the velocity. The behavior of the particle's inertons can be studied in the framework of the Lagrangian (Krasnoholovets 2008).

$$L = -m_0 c^2 \left\{ \frac{T^2}{2m_0^2} \dot{m}^2 + \frac{T^2}{2\Lambda^2} \dot{\vec{\xi}}^2 - \frac{T}{m_0} \dot{m} \nabla \vec{\xi} \right\}^{\frac{1}{2}}. \qquad (21)$$

Here, T is the time period of collisions of the particle and its inerton cloud; $m(\vec{r}, t)$ is the current mass of the {particle-inerton cloud} system; $\vec{\xi}(\vec{r}, t)$ is the variable that describes a local distortion of the tessel-lattice, which can be called a tension (or rugosity). In other words, we introduce a new variable $\vec{\xi}$ for the particle, which means that the mass is not an absolute value but variable. A variation in mass should occur in case of a moving particle (de Broglie 1967).

A similar behavior occurs in the case of a moving electric charge, when it periodically is transferred to the magnetic monopole state in each section of the particle's de Broglie's wavelength (Krasnoholovets2003).

The Euler-Lagrange equations for variables m and $\vec{\xi}$ are

$$\frac{\partial^2 m}{\partial t^2} - \frac{m_0}{T} \nabla \dot{\vec{\xi}} = 0, \qquad (22)$$

$$\frac{\partial^2 \vec{\xi}}{\partial t^2} - \frac{\Lambda^2}{m_0 T} \nabla \dot{m} = 0. \qquad (23)$$

For the system, which features the radial symmetry, variables m and $\vec{\xi}$ are functions of only the distance r from the particle and the proper time t of the {particle + inerton cloud}-system. In this case we preserve only radial components in both variables, which enables us to rewrite equations (22) and (23) in the spherical coordinates as follows

$$\partial^2 m / \partial t^2 - (c^2 / r)\, \partial^2 (r m) / \partial r^2 = 0, \qquad (24)$$

$$\partial^2 \xi / \partial t^2 - (\Lambda^2 / m_0 T)\, (\partial / \partial r)\, \partial m / \partial t = 0 \qquad (25)$$

where the Laplace operator Δ is presented in the spherical coordinates as $\Delta m = \frac{1}{r}\frac{\partial^2}{\partial r^2}(rm)$.

Thus the availability of the radial symmetry and the simple boundary and initial conditions (Krasnoholovets 2008) allow the solutions to equations (24) and (25) in the form of standing spherical waves, which exhibit the dependence $1/r$,

$$m(r,\ t) = C_1 \frac{m_0}{r} \cos\frac{\pi r}{2\Lambda} \left| \cos\frac{\pi t}{2T} \right|, \tag{26}$$

$$\xi(r,\ t) = C_2 \frac{\xi_{\max}}{r} \sin\frac{\pi r}{2\Lambda}\ (-1)^{[t/T]} \sin\frac{\pi t}{2T}. \tag{27}$$

These solutions exhibit the dependence $1/r$, which is typical for standing spherical waves.

The solution for the distribution of mass (26) shows that at a distance $r << \Lambda$ the time averaged distribution of mass of inertons along the radial ray, which originates from the particle, becomes

$$m \approx l_{\text{Planck}} \frac{m_0}{r} \tag{28}$$

In this region the tension of space, as followed from the solution (27), is $\xi \approx 0$.

In such a manner in the space around the particulate cell a peculiar relief is formed, a deformation potential $\propto 1/r$ that spreads up to a distance of $r = \Lambda$. In this range cells of thetessel-lattice are in the contraction state and it is this state of space, which is responsible for the phenomenon of the gravitational attraction. In fact from the distribution (28) we may easily obtain the Newton's gravitational potential

$$u = -G\frac{m_0}{r} \tag{29}$$

where the Newtonian constant of gravitation G plays the role of a dimensional constant.

We may anticipate that a similar pattern should take place around a massive object whose gravity is induced by all its entities. Indeed, entities vibrate in the neighborhood of their equilibrium positions. This occurs in the solid, liquid and gas phases, although in the latter two phases oscillations of entities near equilibrium positions are not stable, and in the gas phase resemble thermal chaotic motion. In any case amplitudes of these oscillations are nothing but the entities' de Broglie wavelengths.

It should be emphasized that the behavior of the speed of an oscillating entity is exactly the same as shown in Figure 1 for the case of a rectilinearly moving particle. Hence vibrating entities of the massive object oscillating in the tessel-lattice produce inerton clouds, which of course overlap. Due to the overlapping a total inerton cloud of the object is formed. The spectrum of inertons is similar to the spectrum of vibrating entities. For instance, if we have a solid sphere with a radius R_{sph}, which consists of N_{sph} atoms, the spectrum of acoustic waves consists of $N_{\text{sph}}/2$ waves with the wavelengths $\lambda_n = 2an$ where a is the lattice constant (i.e. mid-distance between nearest atoms) and $n = 1, 2, 3, ..., N_{\text{sph}}/2$. Note if the lattice constant in a solid is equal to circa 4×10^{-10} m, the lattice constant for protons in the Sun has the same order, about 1.5×10^{-10} m.

At the same time inertons, which accompany acoustically vibrating atoms, produce also their own spectrum and the wavelengths of these collective inertonic vibrations can be estimated by expression

$$\Lambda_n = 2anc/\upsilon_{\text{sound}}, \tag{30}$$

which is the consequence of the relationship (15).

Also note that the behavior of these collective inerton oscillations obeys the law of standing spherical waves, i.e. the dependence of the front of the inerton wave must be proportional to the inverse distance from the source irradiating the wave, $1/r$.For instance, a solid sphere with volume 1 cm^3includes around 10^{22}atoms; putting the velocity of sound $\upsilon_{\text{sound}} \approx 10^3$ m/s and the distance between atoms 0.5 nm, we obtain for the amplitude of the longest inerton wave: $\Lambda_{N/2} \sim 10^{18}$ m. Therefore up to this distance the inerton field of the solid sphere is able to propagate in the form of the standing spherical inerton wave. We may now apply the same consideration to the solid sphere studied, which has been done above for the gravity of a particle. In particular, expressions (26) and (29) is also applicable for the case of a massive object; at distance $r << \Lambda_{N/2}$, which for the solid sphere of volume 1 cm^3 is still a cosmic distance. It is interesting that light takes about 2 years to cover the distance $\Lambda_{N/2} \sim 10^{18}$ m. Note in the case of a material ball with the radius 10 m, its standing inerton waves reach the boundary of the universe, which is estimated by the radius of about 10^{26} m.

At the same time, as has recently been found by the author (Krasnoholovets 2013a), that short wavelength vibrations of atoms in a solid are responsible for the so-called Casimir effect. In this effect associated with the edge of the Brillouin zone, the gravitational potential depends on the distance as $1/r^3$ and short inertonic waves spread only to a distance of r equal to $\Lambda_1 = 2ac/\upsilon_{\text{sound}} \le 1\,\mu\text{m}$.

Thus we are able to derive Newton's gravitational potential (29) also for a macroscopic object in terms of short-range action provided by inertons, carriers of mass entities of objects. If averaged in time, a mass field around the object studied can be considered as a stationary gravitational potential.

Now let us consider the gravitational interaction between two objects, taking Poincaré (1906) note into account that an expression for the gravitation should include the velocity of the attractive object. Recently this has been accented again, i.e. that the law of gravitation has to take into account of the relative speeds of moving masses (Guy 2010).

The sub-microscopic approach (Krasnoholovets 2009) points to the fact that the gravitational interaction between objects must consist of two terms: (i) the radial inerton interaction between masses *M* and *m*, which results in the classical Newton gravitational force

$$F_{\text{Newton}} = G\frac{Mm}{r^2}, \qquad (31)$$

and (ii) the tangential inerton interaction between the central attracting mass *M* and the rotating attractable mass *m*, which is specified by the tangential component of the motion of the test mass *m*.

Indeed, components of the inerton cloud's velocity in the vicinity of the particle, which moves with the velocity $\vec{\upsilon}$, are: $\vec{\upsilon}$ along the particle path and $\vec{c}$ in the transversal directions (Figure 3). The same should be valid for a moving macroscopic object, because individual inerton clouds of vibrating entities in the object overlap forming a total inerton cloud. In the total inerton cloud inertons migrate by the same rule as is the case for inertons of a separate particle: move far away of the object and then return back to it.

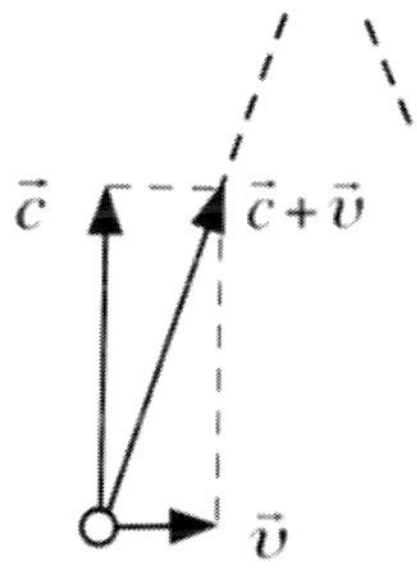

Figure 3. Components the particle's inerton velocity: $\vec{\upsilon}$ is along the particle path, $\vec{c}$ is in the direction transversal to the particle path.

Let a satellite with a mass m enveloped in its total inerton cloud rotate around the attracting mass M. The inerton cloud of the orbital mass m touches the central mass M is partly absorbed by it, which results in the reciprocal interaction between masses M and m. Components of the velocity of this total inerton cloud are $\vec{c}$ along the radial line and $\vec{\upsilon}_{\perp}$ in the tangential direction (Figure 4). Hence the total inerton velocity in the satellite's inerton cloud is

$$\hat{c} = \sqrt{c^2 + \upsilon_{\perp}^2} \tag{32}$$

Then the kinetic energy of these inertons is $mc^2 \cdot (1 + \upsilon_{\perp}^2 / c^2)$.

If the speed of inertons is only c, than the square of a liner element should be written in line with the Minkowski metric,

$$ds^2 = c^2 dt^2 - d\vec{r}^{\,2}. \tag{33}$$

However, since the speed $\hat{c}$ (32) is present in the motion of the particle's inertons, which exceeds the speed of light, the square of a linear element should look as follows

$$ds'^2 = c^2 dt'^2 - d\vec{r}'^2. \tag{34}$$

The time t is the proper time of the particle and t' is the proper time of the particle's inertons. Since these inertons travel faster than c, the relationship between these two times has to be as follows: $\sqrt{1+\upsilon_\perp^2/c^2}\,dt' = dt$. Then equating the linear elements ds^2 and ds'^2 to zero we derive an equation

$$dr'^2 = dr^2/(1+\upsilon_\perp^2/c^2). \tag{35}$$

Now we can solve the problem of how to introduce the tangential velocity into the Newton expression (31) for the radial attraction of two gravitating masses m and M (Figure 4). The real trajectories of inertons of the orbital mass m do not go along the radial line r but have a small shift ahead following the velocity $\vec{c}+\vec{\upsilon}$ of inertons of the orbital mass m; in other words, their trajectory is going along the line r'. Integrating right and left hand sides of eq. (35) we get

$$r'^2 = r^2/(1+\upsilon_\perp^2/c^2). \tag{36}$$

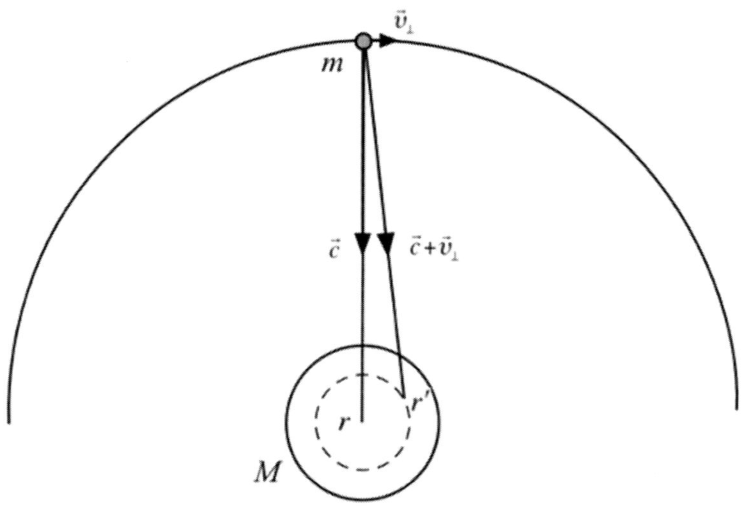

Figure 4. Orbital motion of the mass m around the central massive object with the mass M. Orthogonal components of the velocity of the inerton cloud of the orbital mass are equal to $\vec{c}$ and $\vec{\upsilon}_\perp$. Due to the tangential component $\vec{\upsilon}_\perp$ the line of attraction between the masses M and m shifts from the centre of the mass M and the effective radius r' of the attraction becomes shorter than r.

Substituting in expression (31) r^2 for r'^2 from expression (36) we obtain

$$F = G\frac{Mm}{r^2}(1 + \upsilon_\perp^2 / c^2). \tag{37}$$

Then for the gravitational interaction of the orbital mass m with the central mass M we have

$$U = -G\frac{Mm}{r}\left(1 + \frac{r^2\dot{\varphi}^2}{c^2}\right), \tag{38}$$

where we put for the tangential velocity $\upsilon_\perp = r\dot{\varphi}$.

Expression (38) was successfully used to obtain exactly the same equations and the solutions for such problems as the motion of Mercury's perihelion, the deflection of starlight by the Sun, and the gravitational redshift of spectral lines (Krasnoholovets 2009). Besides, this approach allowed the derivation of the Shapiro time delay effect (Krasnoholovets 2013b).

In contrast to orthodox quantum theory and general relativity, the submicroscopic concept allows us to derive the Newton gravitational potential (29) and introduce the corrected version of Newton's law of gravitation (38).

Our research has shown that the Schwarzschild-Hilbert metric (4), which represents properties of the geometry of space-time of a point mass M at rest, implicitly includes the second term of physical interaction associated with the orbital motion of a test mass m in the location of the latter (38).

Fractality of the tessel-lattice and fractality of balls, which compose the tessel-lattice, are responsible for the real geometry of physical space. This means that space-time of general relativity is disclosed as the four dimensional space of the tessel-lattice. In the tessel-lattice a convolution product represents the fundamental metrics of the ordinary physical space where the embedding part allows the description by the following relation (Bounias and Krasnoholovets 2003b)

$$\mathrm{D4} = \int\left(\int_{dV}(d\vec{x}\cdot d\vec{y}\cdot d\vec{z})\right) * d\Psi(w) \tag{39}$$

where dV is an element of space, and $d\Psi(w)$ a function accounting for the extension of coordinates to the 4th dimension through convolution ($*$) with the volume of space.

The fourth dimension reflects the space fractality, i.e. fractality of the tessel-lattice's balls. Time determined as a natural parameter through the path, can change only when balls, which form the path, shrink. Therefore non-linear components of metric of general relativity shall be considered as a mapping of original shrunk balls of the tessel-lattice (Krasnoholovets 2013b).

3. Cosmology and Dark Energy from the Submicroscopic Point of View

3.1. The Modern Paradigm and its Problems

Cosmology is the science about the universe in general, the most general laws of the universe formation, structure and evolution. General laws of largest-scale structures and dynamics of the universe are studied by means of cosmological models. Modern cosmology is based on Einstein's general relativity, especially the Friedmann model of the expansion of the universe (Friedmann 1922). The metric in the corresponding Einstein field equations is the Friedmann–Lemaître–Robertson–Walker metric (Friedmann 1922; Lemaître 1931; Robertson 1935; Walker 1937); it describes a homogeneous, isotropic expanding universe.

A very popular theme in cosmology is the beginning of the universe, which allegedly started from the Big Bang. The idea was formed soon after first works of Friedmann and Lemaître on the model of the expanding universe (Lemaître 1931).

In modern literature the Friedmann equations look as follows

$$\frac{\dot{a}^2 + kc^2}{a^2} = \frac{8\pi G\rho + \Lambda c^2}{3}, \tag{40}$$

$$\frac{\ddot{a}}{a} = -\frac{4\pi G}{3}\left(\rho + \frac{3P}{c^2}\right) + \frac{\Lambda c^2}{3}, \tag{41}$$

where the parameter k takes the values 0, 1, -1 and determines whether the shape of the universe if flat, a closed sphere or an open 3-hyperboloid; $a(t)$ is the scale factor; $H = \dot{a}/a$ is the Hubble parameter; Λ is the cosmological constant; ρ is the density of matter; P is the internal pressure. In cosmology an important parameter is the density parameter defined as

$$\Omega \equiv \frac{\rho}{\rho_c} = \frac{8\pi G\rho}{3H^2}, \tag{42}$$

where ρ_c is the critical density. The total density parameter is subdivided to a series of components:

$$\Omega = \Omega_{\mathrm{m}} + \Omega_{\mathrm{rel}} + \Omega_{\Lambda}, \tag{43}$$

where Ω_{m} is the mass density including baryonic mass and dark matter, Ω_{rel} is the effective mass density of relativistic particles (light and neutrinos), Ω_{Λ} is the effective mass density of the dark energy related to the cosmological constant Λ .The standard model of cosmology focuses on measurement and determination of these cosmological parameters (Freedman 2000).

On the other hand, the Friedmann model and the related Big Bang theory meet severe difficulties. The first one is the problem of singularity. The solution leads to the conclusion that in the initial moment $t = 0$ the radius of the metagalaxy was zero and hence the density of matter in that moment equaled infinity. Such singularity contradicts the whole physical experience.

The second one is the problem of horizon. It is believed that the microwave background radiation comes from widely separated parts of the universe. However, these regions are so separated that they cannot even communicate by means of signals travelling with the speed of light. The question is: why do these regions have exactly the same temperature 2.7 K? If in the past the metagalaxy was divided into many causally non-related regions, then how can we resolve a problem associated with the observed amazing isotropy of the metagalaxy? How could different parts of the metagalaxy causally unrelated adjust to each other, so that by now a perfect isotropic (spherical or quasi-spherical) geometry has emerged? To resolve this paradox, the Big Bang theory simply assumes the needed level of uniformity.

The third is the flatness problem. The observations point out that the universe has very low geometrical curvature in its space-time (it is nearly flat). So logically it is very unlikely that the universe indeed started from the Big Bang.

The forth is the monopole problem. Grand unified theories of physical interactions predict the existence of very massive particles called magnetic monopoles that should have been created at the Big Bang. However, these particles still have not been discovered. So, the problem is: where are the monopoles?

Moreover, some researchers formed a list of 30 problems associated with the Big Bang (The Top 30 Problems with the Big Bang, 2002)

Researchers try to solve these and smaller problems by introducing different novelties in the existing concepts. However, there is a possibility to remove these difficulties in a revolutionary way by changing the paradigm.

In fact, general relativity is a local theory written for an abstract massive point, which by the author's estimation cannot be less than the size 10 μm, because at a smaller size the laws of quantum physics start. At larger sizes the inner structure of matter has to influence the formalism of general relativity through vibrations of the matter's entities, which brings the matter's cloud of inertons into play.

The Friedmann model looks very artificial itself: 1) a local theory was applied to the whole universe; 2) it is based on a non-physical metric (the so-called the Friedmann–Lemaître–Robertson–Walker metric). Friedmann (1922) was the first who introduced time into spatial components of the metric

$$ds^2 = R^2(x_4)(dx_1^2 + \sin^2 x_1\, dx_2^2 + \sin^2 x_1 \sin^2 x_2 dx_3^2) + M^2(x_1, x_2, x_3, x_4) dx_4^2 \quad (44)$$

where the coordinate x_4 is associated with the time t; $M^2(x_1, x_2, x_3, x_4)$ is a function (it is not a mass); $R(x_4)$ is proportional to the radius of curvature of space, so that the radius of curvature R can change over time. Today instead of Friedmann's designation $R(x_4)$ the curvature of space is written as $a(t)$ and called a scale factor. The Friedmann equations are derived from the Einstein-Hilbert equations (1) by using the metric (45).

However, in the Riemannian geometry, the scalar curvature (or the Ricci scalar) R is the simplest curvature invariant (scalar) of a Riemannian manifold. Its geometrical meaning is reduced to a single real number determined by the intrinsic geometry of the manifold near the point in which R is calculated. The scalar curvature R represents the difference in volumes of a geodesic ball written in a curved Riemannian manifold and the ball in Euclidean space. Thus R is a number, for example, 3.14. According to Friedmann, this number can depend on time t, namely, 3.14(t). Substituting in the metric (40) the scalar factor $R \equiv \text{const}$ with a factor depending of the time variable, $R(t) \equiv 3.14(t)$, one may try to solve the Einstein-Hilbert equations (1). However, how can such an approach be related to physics??

The cosmic microwave background, which is very important for the standard model of cosmology, seems a fiction (Robitaille 2007; 2009; 2010a; 2010b).

So, in physics there is no place for the Friedmann equations, the Big Bang and the cosmic microwave background. Then what are the origin of dark energy, the Hubble constant and other cosmological parameters? To understand the origin, researchers should go through an inner phase transition in their mind. Namely, they have to stop to consider themselves as ichthyologists who study the behavior of fish (i.e. stars and galaxies), but rather oceanographers who study the ocean as the whole (the universe that is a peculiar substrate in which stars and galaxies are excitations that behave like fish in the ocean).

3.2. The Universe as a Huge Cluster in the Tessel-Lattice

The cellular structure of superclusters of galaxies and a sphere-like structure in aggregations of galaxies of smaller sizes are two problems of astrophysics. Any theory that attempts to explain the origin of large-scale structure of the universe must address these two problems.

Studies of relative scales in the empty-set lattice showed that the antifounded properties of the empty set provide existence of a lattice involving a tessellation of the corresponding abstract space with empty balls, i.e. the tessel-lattice (Bounias and Krasnoholovets 2003a, 2003b).Properties of a space at micro-scale are provided by properties of the spaces whose members are empty set units. In the mentioned works it was shown that particular levels of a measure of these units, quantum levels at relative scales, can be discerned.

A finite set of rational numbers inferring from a Cartesian product of a finite beginning section of integer numbers establishes a discrete scale of relative sizes. One possibility of a scaling progression covering integer subdivisions (n) consists in dividing a fundamental segment (n = 1) by 2, then each subsegment by 3, etc. Hence the size of structures is a function of iterations (n). At each step $(\nu_{\rm j})$ the ratio of the size in the dimension D is $(\prod \nu_{\rm j})^{\rm D}$, so that the maximum ratio becomes

$$\Re \propto \left\{ \left(\prod \nu_{\rm j} \right)^{\rm D} \left(\prod \nu_{\rm j} - 1 \right) \right\}_{\rm j=1 \to n} . \tag{45}$$

The manifold $(\prod \nu_{\rm j})^{\rm D}$ is a commutative Bourbaki-multipliable indexed on the integer section $\rm I = [1, n]$.

Expression (45) is of interested to us, as it allows one to establish a range and intermediate levels of the scale of size of objects composing a universe (Bounias and Krasnoholovets 2003b). Set theory and fractal geometry predict orders of size, different levels of organization (how many structural units this or that scale level requires in the universe). Predictable orders starting from the Planck scale, roughly comply with quark-like size (10^{10} to 10^{11} units), particle to atoms (10^{11} to 10^{17} units), molecules (10^{21}), human size (10^{28}), stars and solar systems (10^{40} to 10^{42}), up to the estimated upper limit (10^{56}) that could be bounded by a scale at (10^{60} to 10^{61}).There should be further levels of higher scale universes, kinds of continued clusters with the following numbers of units: from $10^{82}$to 10^{120}, then 10^{139} and 10^{142} to 10^{171}.

In the microworld, an introduction of an empty hyperset ∅ allowed us the implementation of almost countless number of iterations in a separate cell of the tessel-lattice. As we just showed above, the tessel-lattice allows different levels of self-organizations at high scales. In particular, it predicts the existence of galaxies, their unification and also the formation of a set of universes as well. The world of universes might look as depicted in Figure 5.

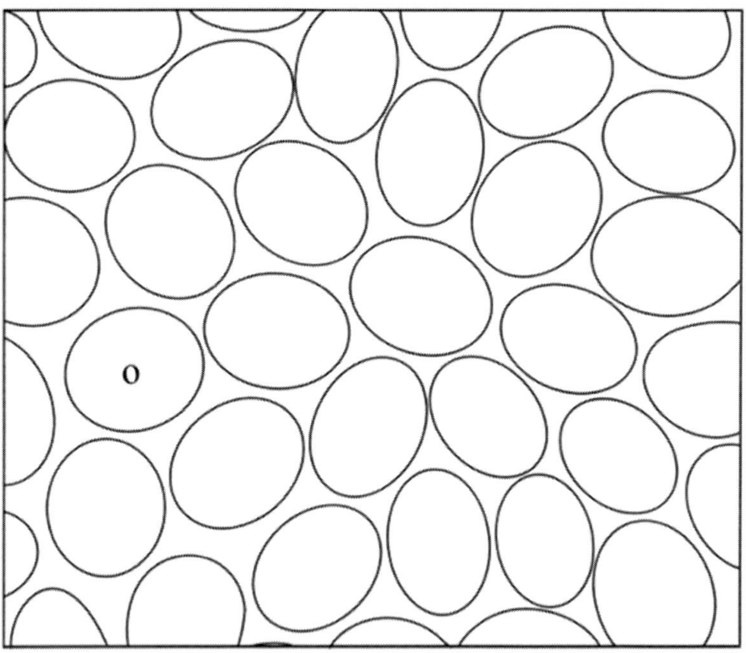

Figure 5. Manifold of spheroidal universes. The universe marked with the letter "O" would be ours. This is the pattern of one of the levels of higher scale universes, for examples, it can be the biggest cluster structure in the total world tessel-lattice.

Although the total space is subdivided into clusters and universes, the tessel-lattice should remain uninterrupted. This can be obtained through the different states of lattice balls inside universes and between them. Inside a universe they must be more contracted than outside the universe, which should support the tessel-lattice in a balanced state. In Figure 6 such balls are shown for these two regions. A pair of charged particles emerges from a couple of balls in a universe. Leptons are strongly fractally deformed and quarks are maximally inflated. The charge state is determined by a polarization of the surface of a particulate ball (Krasnoholovets 2003).

What kind of organization, or ordering do topological balls experience in a spheroidal universe? To determine this, we may compare the problem with a macroscopic ball formed by a fluidic matter that shares also properties of gaseous and solid phases. Since the universe possesses a spherical shape, it should be characterized by the Laplace pressure $\gamma_{\text{univ}} \cdot (1/R_{\text{univ}1} - 1/R_{\text{univ}2})$ that forms a shell around the body of the universe. Hence the total pressure inside of the universe can vary from the center to the periphery. Namely, the size of a ball in the tessel-lattice may gradually increase from the center to the periphery (Bounias and Krasnoholovets 2003c); the radius of a ball in the *n*th coat can be approximated as follows:

$$\ell_n = \ell_0 + (n-1)b \tag{46}$$

where ℓ_0 is the minimal size of the ball at the center of the universe and b is a tiny increment of ℓ_0. In the continuous approximation the ratio ℓ_n / ℓ_0 (46) can be rewritten as $1 + x$ where $x = r/R_{\text{univ}}$; here, r is the radial distance from the center of the universe and R_{univ} is the radius of the universe.

Since we admit the variation of the ball size, we have to permit a variation in the pressure of the universe too, because the surface of balls changes towards the periphery of the universe. Since we allow the linear rule (43) for increasing the radius of a ball, the pressure at a point, which is removed at a distance r from the center of the universe, can be written as

$$P(x) = P_0 / (1+x)^2 \tag{47}$$

where P_0 is the pressure in the center of the universe; $x = r / R_{\text{univ}}$ is the dimensionless variable in which R_{univ} is the radius of the universe estimated by astronomers as equal to around 10^{26} m.

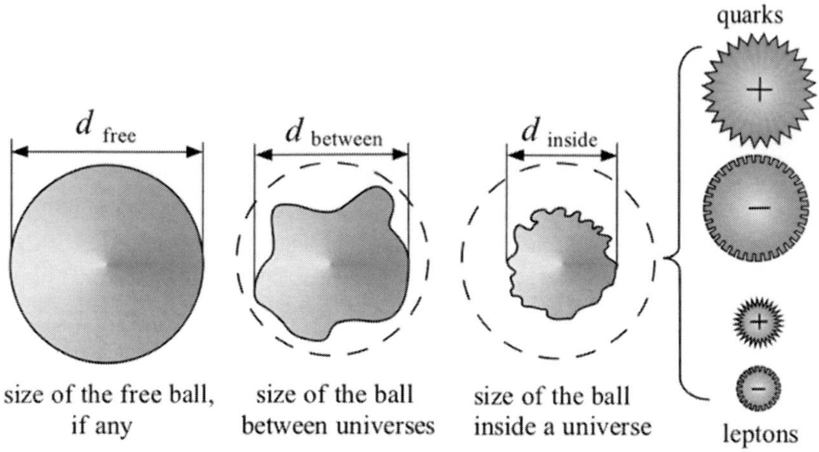

Figure 6. Topological ball, a primary subtle particle. Due to fractal deformations the ball shrinks, which is most significant insides a universe than in the interior regions.

The mass of a ball inside the universe is defined as the ratio of its fractal volume to the middle volume of a ball from the region between universes (compare with expression (8)); such a "zero"-mass is the reference point for real masses (baryons) in the universe. The solution (26) for the distribution of mass around the massive object hints to a distribution of the density of topological balls inside the universe. This distribution should be a stable stationary solution for the cluster of a larger scale in the tessel-lattice, which is the universe. So, the distribution of the density of the tessel-lattice in the universe can have the form of a standing spherical wave

$$\rho(x) = \rho_0 \frac{\cos x}{1 + x}, \tag{48}$$

where ρ_0 is the density of the tessel-lattice in the central part of the spherical universe. It is obvious that x has to vary from l_{Planck} to $\pi R_{\text{univ}} / 2$, such that at the boundary of the universe $\rho_{\text{bound. univ}} = 0$.

Formulas (47) and (48) show that remote distances can be overestimated by a measure using a local gauge.

Expressions (47) and (48) allow us to calculate the finite sound velocity in the universe as the function of a distance x from its centre. In fact,

$$\frac{\partial P}{\partial \rho} = c^2, \tag{49}$$

where the derivative should be taken at the constant entropy, which is natural for the tessel-lattice. The equation (49) can be rewritten as follows

$$\frac{\partial P}{\partial x} \cdot \frac{1}{\frac{\partial \rho}{\partial x}} = c^2, \tag{50}$$

which results in

$$c(x) = \sqrt{\frac{P_0}{\rho_0}} \frac{1}{\sqrt{(1+x)^2 \sin x + (1+x)\cos x}}\Bigg|_{x \ll 1} \cong \frac{\sqrt{P_0 / \rho_0}}{1 + x + x^2}, \tag{51}$$

or finally putting $c \equiv \sqrt{P_0 / \rho_0}$ (the speed of light in the central part of the universe) and retaining only linear members by $x \equiv r / R_{\text{univ}}$ we get instead of expression (51)

$$c(r) \cong c \cdot \frac{1}{1 + r / R_{\text{univ}}} \tag{52}$$

It should be noted that in the tessel-lattice two kinds of basic deformations of a cell exist: a fractal volumetric deformation, which is associated with the physical notion of mass, and a surface fractal deformation, which is associated with the physical notion of electric charge (Krasnoholovets 2003). That is why two sound velocities exist in the tessel-lattice: fast longitudinal c_{inerton} and slower transversal c_{photon}. The general formulas should be the same, (51) and (52), in which, however, the density has a different nature – volumetric (mass) and surface (charge). In other words, for each ball there should be $\rho_{0\,\text{volume}}$ (responsible for mass-tension transitions) and $\rho_{0\,\text{surface}}$ (responsible for charge-monopole transitions). What is more, the inequality $\rho_{0\,\text{volume}} << \rho_{0\,\text{surface}}$ should hold, because, as is known from condensed matter physics, the longitudinal velocity exceeds the transversal one, which for our case means that photons are slower than free inertons, $c_{\text{photon}} << c_{\text{inerton}}$.

Below we will discuss expression (52) implying that it is recorded for photons, i.e., c is the speed of light in the center of the universe, and $c(r)$ is the speed of light at a distance r from the center towards the periphery.

Let us now consider Hubble's law, which is expressed by the equation $\upsilon = H_0 D$ where H_0 is the Hubble constant and D is a distance; $\upsilon$is the recessional velocity, in km·s^{-1} (Hubble 1929).The parameters H_0 and $D$are not directly measured. They are found through a supernova brightness, which provides data on its distance, and the redshift $z = \Delta\lambda/\lambda$ of its spectrum of radiation (the fainter and smaller a galaxy appears, the higher is its redshift, which typically is due to the Doppler effect). Hubble tied the brightness and redshift parameter z suggesting a surprising correlation

$$H_0 D = \frac{\lambda_{\text{measured}} - \lambda_{\text{source}}}{\lambda_{\text{source}}} c\,, \tag{53}$$

which suggests that more distant galaxies are moving faster away from us.

Thus by the relationship (53) the more distant galaxy is, the faster it is moving away from us, which means that the recessional velocity of a galaxy is proportional to its distance from us, $\upsilon = H_0 D$. The relationship (53) has been proved experimentally and is considered now as the major argument supporting the expansion of our universe.

However, let us rewrite the relationship (53) in forms

$$\lambda_{\text{source}} = \frac{\lambda_{\text{measured}}}{1 + H_0 D/c}, c/\nu_{\text{source}} = \frac{c/\nu_{\text{measured}}}{1 + H_0 D/c}. \tag{54}$$

From the second version of expressions (54) we can see that the other explanation of the phenomena of the recessional velocity is also possible. Namely, we may shift the change in the frequency, $\nu_{\text{source}} \to \nu_{\text{measured}}$, to the change in the speed of light: at a distance D from a terrestrial observatory the velocity of light can be different from the value of c measured in the central part of the universe. The appropriate relationship becomes

$$c(D) = c \cdot \frac{1}{1 + H_0 D / c}. \tag{55}$$

It is easy to see that the relationship (55) can be related to expression (52). In fact, let us check the equality $1/R_{\text{univ}} = H_0 / c$. Putting the Hubble constant $H_0 = 70$ km·s^{-1}·Mpc $= 2.265 \times 10^{-18}$ s^{-1} and the speed of light $c = 3 \times 10^8$ m·s^{-1}, we get for the radius of the universe $R_{\text{univ}} = 1.32 \times 10^{26}$ m, which is in agreement with the modern data. Hence the tessel-lattice model of cosmology naturally interprets the Hubble's law as a gradual decrease of the speed of light towards the periphery of the universe. No redshift, but instead the loss of speed of light owing to the step-by-step reduction of the density of the tessel-lattice in the framework of the universe. This also means that there is no dark energy in the universe, which allegedly accelerates galaxies after the hypothetical Big Bang.

The problem of the so-called accelerating expansion of the universe with high-redshift supernovae $z \approx 0.5$ (Perlmutter et al. 1998; Schmidt et al. 1998) has a natural elucidation in the framework of the present submicroscopic concept. The velocity of light is a function of the distance r from the center of the universe and the velocity $c(r)$ gradually decreases and especially dramatically drops near the boundary of the universe reaching the minimum value $0.39c$ at the boundary (see expression (51)). This means that the observed “accelerating expansion of the universe” can be accounted for by an abrupt dropping of the speed of photons in the vicinity of the boundary of the universe where the 42 Type Ia supernovae were detected. In fact, the derivative of the speed $c(r)$ (51) and (52) with respect to distance r is negative, which means a deceleration of photons at that point. However, if one does not take into account an abrupt reduction of the speed of light near the periphery, the supernovae could be perceived by an observer on the Earth to be accelerated in the direction from the center of the universe.

4. Dark Matter

Dark matter is one more puzzle of modern physics, astrophysics and astronomy. The source of the problem relates to the perception of the universe as an empty reservoir filled with residuals of the Big Bang. However, the consideration of orderly physical space as the tessel-lattice removes all the difficulties in the description and comprehension of this phenomenon (Krasnoholovets 2011).

Stars are found in the tessel-lattice and their inner entities vibrate near their equilibrium positions. Due to the interaction with cells of the tessel-lattice the entities emit inertons, which form the total inerton cloud of the star inducing the Newton potential (29). This inerton cloud is distributed around the star as a standing inerton wave, which spreads up to the boundary of the universe. In an ensemble of stars a mutual overlapping of the stars' inerton clouds occurs. Standing inerton waves of the opposite sources should undergo an elastic interaction. For instance, a similar picture we can observe on the surface of water when water waves of two sources interfere, generating wavelets.

Therefore, standing inerton waves induce the gravitational potential (29) and, in addition, owing to the mutual scattering of counter propagating waves of nearest stars they introduce the elasticity in interstellar space.

The Hamiltonian of interacting stars can be written in a lattice model in the form

$$H(n) = \sum_{s} E_s n_s - \tfrac{1}{2}\sum_{s,s'} u_{ss'}^{\text{attract}}\, n_s n_{s'} + \tfrac{1}{2}\sum_{s,s'} u_{ss'}^{\text{repul}}\, n_s n_{s'}. \qquad (56)$$

Here E_s is the additive part of the star energy (the kinetic energy) in the sth state. Then we have two kinds of pair potentials: the attraction and the repulsion components. So, in the Hamiltonian (56) the potential $u_{ss'}^{\text{attract}}$ represents the paired energy of attraction and the potential $u_{ss'}^{\text{repul}}$ is the paired energy of repulsion. All starts are considered as identical. The potentials take into account the effective paired interaction between stars located in states s and s'. The filling numbers n_s can get only two values in the model lattice: 1 (the sth knot is occupied) or 0 (the sth knot is not occupied). The signs before positive functions $u_{ss'}^{\text{attract}}$ and $u_{ss'}^{\text{repul}}$ in the Hamiltonian (56) directly specify proper signs of attraction (minus) and repulsion (plus).

The statistical sum of the system of interacting stars

$$Z = \sum_{\{n\}} \exp\left(-H(n)/k_{\text{B}}\Theta\right) \qquad (57)$$

can be rewritten via the action S, which depends on three functions,

$$Z = \mathrm{Re}\frac{1}{2\pi i}\int D\phi \int D\psi \oint dz \exp\left[S(\phi,\ \psi,\ z)\right]. \tag{58}$$

The complicated function $S(\phi,\ \psi,\ z)$ was evaluated for extremum (Krasnoholovets and Lev 2003). The minimum of the action S is reached when all stars are distributed by clusters, especially if each cluster includes the same number of stars. In this case the action for a cluster of N stars (allowing they obey the Bose statistics) becomes

$$S \approx \tfrac{1}{2}\{a(N) - b(N)\} \cdot N^2 \tag{59}$$

where the functions a and b are defined as follows:

$$a(N) = 3\int_1^{N^{1/3}} u^{\text{repul}}(gx)\, x^2\, dx/(k_{\text{B}}\Theta), \tag{60}$$

$$b(N) = 3\int_1^{N^{1/3}} u^{\text{attract}}(gx)\, x^2\, dx/(k_{\text{B}}\Theta) \tag{61}$$

In the case of the Boltzmann statistics the action for a cluster of N stars has the form

$$S = 2\{N\cdot [a(N) - b(N)] + N\ln \xi_0\} \tag{62}$$

where ξ_0 is the fugacity.

Knowing the explicit form of the action (59) or (62), one can derive the equation for the number of particles combined in a cluster: $\partial S/\partial N = 0$, which in addition requires holding of the inequality $\partial^2 S/\partial N^2\,|_{N=N_{\text{in cluster}}} > 0$.

The pair potentials that enter equations (59)-(62) look as follows

$$u^{\text{attract}}(gx) = \frac{Gm^2}{gx} + \frac{GMm}{R}, \tag{63}$$

$$u^{\text{elast. repul}}(gx) = \tfrac{1}{2}m\omega^2 (gx)^2 \tag{64}$$

where m is the mass of a star; g is the lattice constant, the distance between neighbor stars in the model lattice); x is the dimensionless distance defined through a relation $r = gx$; ω is the radial frequency of oscillations of the mass m near its equilibrium position. Besides, in expression (63) we add a mean field formed by the whole galaxy, where M is the total mass of the whole system of stars and R is the effective radius of the system.

It should be noted that a quadratic potential $V(r) = \gamma_0 c^2 r^2 / 2$ was already used in the conformal theory of gravity to fit the observed behavior of galactic rotation curves (Mannheim and O'Brien 2010, 2011). The origin of the potential generated through the effect of cosmology on individual galaxies through all parameters had a completely different meaning originating from an abstract metric of a theory of long-range action despite its quantization.

The solutions for clusters of stars can be obtained for all the dimensions: 3D, 2D and 1D (Krasnoholovets 2011). All three cases were observed: bulk galaxies (Clowe et al. 2004), disc galaxies (Famaey and McGaugh 2013) and rings/arcs (Fort and Mellier 1994; Jee et al. 2007), respectively.

The most interesting are solutions for disc galaxies and bulk galaxies. Starting from the action (59), we get for a disc galaxy the following solution

$$N = \frac{3}{2} \frac{GM}{Rg^2\omega^2}. \tag{65}$$

which means that all the stars of the disc galaxy are distributed by a plane cluster with the appropriate number N of stars. We may choose some typical values of the mass, the radius and the distance between stars in the galaxy: $M = 10^8 M_{\oplus}$, $R = 10$ kpc, $g = 1$ pc. The frequency ω of oscillations of a star near its equilibrium position in the cluster can be estimated from the equality of the centripetal acceleration that attracts a star to the center of the galaxy and the holding force that acts between stars in the cluster:

$$GM / R^2 - \omega^2 g = 0. \tag{66}$$

From eq. (66) we get

$$\omega = \left(\frac{GM}{R^2 g} \right)^{1/2} \approx 2.1 \times 10^{-14} \ [\mathrm{s}^{-1}]. \tag{67}$$

Let us now evaluate the acceleration that each star experiences in the plane cluster:

$$a = GM / R^2 \approx 1.4 \times 10^{-11} \text{m·s}^{-2}. \quad (68)$$

This value of the acceleration satisfies the conditions prescribed by Milgrom (Famaey and McGaugh 2012, 2013) $a << a_0 \cong 1.21 \times 10^{-10}$ m·s^{-2}. Thus we do not need to assume an incomprehensible modernization of Newton's law, i.e. the substitution of the force $\mathbf{F} = m\mathbf{a}$ by a significantly smaller force of undetermined nature $\mathbf{F} = m \cdot (\mathbf{a} - \Delta\mathbf{a})$ at $a << a_0$. Stars are distributed by clusters and each star is strongly bonded with the other $N-1$ stars in the cluster. This bonding compensates the centripetal acceleration. Expression (66) demonstrates this balance of two competing forces. That is why a Keplerian law, $V \sim 1/\sqrt{r}$, is substituted for Milgrom's constant orbital velocity $V^4 = GMa_0$.

Let us now discuss the 3D cluster solution

$$N = \left(\frac{30}{11} \frac{GM}{Rg^2 \omega^2} \right)^{3/2}. \quad (69)$$

The parameters can be chosen as follows: the total mass $M \sim 10^{14} M_\oplus$, the radius $R \approx 250$ kpc and the central mass density $\rho_0 = 3.85 \times 6 M_\oplus$ kpc^{-3}. Putting for the mass of a star $m = M_\oplus$, we obtain the mean distance between stars: $g = (M_\oplus / \rho_0)^{1/3} = 4.25 \times 10^{16}$ m. Then the stability of the cluster in respect to its gravitational collapse is determined by the relationship (66): the gravitational attraction of stars to the centroid is retained by the elasticity of inerton waves in the cluster. The corresponding frequency of oscillating stars at their equilibrium positions owing to the overlapping of their inerton waves is

$$\omega = \left(\frac{GM}{R^2 g} \right)^{1/2} \approx 3.82 \times 10^{-13} \text{ [s}^{-1}\text{]}. \quad (70)$$

The acceleration to the centroid, which each star experiences in the spherical cluster, is

$$a = GM / R^2 \approx 2.24 \times 10^{-8} \text{m·s}^{-2}. \tag{71}$$

The acceleration (71) is opposite to the inequality $a << a_0$ needed for the use of MOND. Besides, the acceleration (71) is not compensated by the acceleration caused by the elastic interaction in the cluster of N stars: $\omega^2 g \approx 6.2 \times 10^{-9}$ m·s^{-2}.

However, the origin of so-called dark matter is nothing but the same stars, which are involved in the mutual interaction through their inerton waves. Stars are floating in an elastic space. This elastic interaction is responsible also for a non-homogeneous distribution of stars, i.e. they assemble in clusters. Hence in a galaxy there must exist a non-equilibrium density distribution of mass. It is this regularity that has to be responsible for the correct rotation curve (Feng and Gallo 2008, 2011). By introducing a complicated distribution of mass in a disc galaxy Feng and Gallo (2008, 2011) managed to obtain the correct rotational curve profile even without an introduction of dark matter. But where is the origin for a no-homogeneous distribution of mass? The answer is obvious: the elastic interaction affects the density distribution and influences on the rotational curve profile bringing it to a law $\upsilon^4 = GMa_0$.

Basic concepts of gravitational lensing should also be modified – perhaps a pointmass approach with a correction based on MOND or another model will require a substitution by an approach resting on the involvement of elastically interacting masses. In particular, it seems the deflection angle $\varphi = 4Gm/(c^2 r)$ of a point mass m, which includes the absolute value of the gravitational potential Gm/r, can be modified as follows

$$\tilde{\varphi} = \frac{4}{c^2}\left(\frac{Gm}{r} + \tfrac{1}{2}\omega^2 r^2\right); \tag{72}$$

In a cluster the second term in expression (72) tends to align the space deformed by the first term. This has to be typical for 2D clusters (i.e. clusters in disc galaxies). In the case of 3D clusters (rich clusters in galaxies) the second term may even prevail the first one, namely, the second term prolongs the deflection angle φ for larger distances at which the first Newton's term becomes already negligible.

The center of a galaxy that can be described as a cluster of stars possesses a peculiar property. This is a small region in which inerton standing waves of different stars interfere. Inerton waves deliver mass to this place, but the waves do not spread further and therefore the mass m is not transferred to the tension ξ as equations (22) and (23) prescribe. This region looks as an endpoint of standing inerton waves of each star of the galaxy. This is a kind of singular point, which might be observed by astronomers and interpreted as a black hole. Nevertheless, inertons do not disappear at this point but establish a special stationary relief of mass from stars to the endpoint supported by the equilibrium of the centripetal acceleration (generated by the rotating stars) and the centrifugal one (emerged as the linear response to the inerton wave pressure on the side of the elastic tessel-lattice).

The long-standing problem about the additional enigmatic acceleration of artificial satellites in the solar system has been debated for years. In particular, researchers discussed some Doppler shift anomalies for the Pioneers 10 and 11, Galileo, NEAR, Cassini, Stardust, Rosetta, Hayabusa, and MESSENGER, as well as the flyby anomalies (Lämmerzahl et al. 2008).

Based on data for the Pioneers 10 and 11 orbit determination and more detailed studies of all the systematics, a total error budget, i.e. an acceleration pulling the satellite Pioneer 10 to the sun was determined as $a_{\text{Pioneer}} = 8.74 \times 10^{-10}$ m·s^{-2} (Anderson et al. 2002). However, later on the same researchers changed their opinion (Turyshev et al. 2012): they mention that since the spacecraft was facing the Sun, the solar energy was absorbed primarily by the high-gain antenna that largely shadowed the rest of the spacecraft from direct solar irradiation; the absorbed energy increased the high-gain antenna temperature and was then emitted as infrared radiation. A contribution on the side of this thermic radiation emanating from the antenna was estimated of the same value as the Doppler's shift.

However, the hypothesis of the thermal origin of the Pioneer 10 anomaly (Turyshev et al. 2012) has a significant flaw, because the further the distance of the satellite from the Sun, the lower is its heating by the Sun and hence the effect of the thermic radiation of the antenna should also be weaker. Unfortunately, the authors of the hypothesis did not pay any attention to this fact.

Lämmerzahl et al. (2008) also discussed other anomalies, such as the increase of the astronomical unit (the distance between the Earth and the Sun), the quadrupole and octupole anomaly, and dark energy and dark matter.

It seems likely that all the anomalies mentioned above are directly associated with the inerton elasticity of space, which is generated at the gravitational interaction of cosmic objects. In fact, a spacecraft moves through a viscous network formed by standing inerton waves of interacting stars, planets and other cosmic objects. Of course the viscosity of space has to slow the movement of satellites (through the switching of terms $\kappa\dot{r}$, $\frac{1}{2}\gamma r^2$, and the Navier-Stokes equation), which so far has never been taken into account by the staffs that led the spacecrafts. In particular, expressions (68) and (71) allow for the introduction of the viscosity of space in our galaxy related to the Milky Way's "dark matter".

In recent work (Turyshev et al. 2012) a realization of astronomical relativistic reference frames in the Solar System and its application to the GRAIL mission are presented. The authors develop the relevant relativistic coordinate transformations that are needed to describe the motion of the GRAIL spacecraft and to compute all observable quantities. This is made in a general relativistic model (calculations involve an expansion of the metric tensor and the Newtonian potential preserving a few terms in the expansion) with accuracy to 1 μm. They derive phase differences measured at both spacecraft, $n_{AB}(t\mathrm{B})$ and $n_{BA}(t\mathrm{A})$, together with the instantaneous delays measured at the points of signal reception at both spacecraft, $T_{AB}(t\mathrm{B})$ and $T_{BA}(t\mathrm{A})$. It can be argued beforehand that the data received from the satellites will have inaccuracies due to the neglect of the viscosity of space in which the satellite is moving.

5. First Steps of Inerton Astronomy

Non-stability of cosmological physical factors that affected different biophysical, chemical physical and physical processes in a laboratory was studied by Sholl with collaborators (Shnoll et al. 2000, Shnoll et al. 2005; Shnoll 2001, 2009). For example, processing of six-year measurement data of alpha-activity of ^{239}Pu with the total of about 60,000 measurements showed regularity of patterns of histograms (the number of recorded counts per minute), namely, the highest probability of similarity nearest neighbors (the effect of near-field) and a clear circadian periodicity of similar histograms. They also observed a change in timing of histograms constructed by measurements of α-activity of ^{239}Pu on a ship in the Indian Ocean and in the laboratory near Moscow in 1988. A direct correlation with the position of the Moon in the sky (especially at the rising and setting of the Moon) and the form of histograms plotted by data of alpha decay of ^{239}Pu.

A coincidence of histograms was observed, which were recorded by measurements of γ-activity of ^{137}Cs at the Columbus Polarity Therapy Institute in Columbus by M. Sue Benford and J. Talnagi and α-activity of ^{239}Pu by S. Shnoll in Pushchino (Russia) during the period of18-20 February 2001, separated by a distance of 12,000 km.

In the study of one-minute histograms (constructed for each of 60 one-second measurements of alpha activity of ^{210}Po) periodic signal repetitions were detected, "macroscopic fluctuations", which were associated rather with a sidereal day

(1436 hours) but not the solar day (1440 hours).Many months of continuous studies pointed out to periodic cosmic fluctuations in the intensity of fundamental physical processes that manifested themselves each 546 hours (with a period of 22.75 days), 576 hours (with a period of 24 days), 648 (with a period of 27.0 days) and 672 hours (with period of 28 days).

Later on Shnoll and co-workers carried out experiments with collimators directed towards the Polar Star and the Sun, the West and the East, as well as with clockwise and counterclockwise rotations, then with GCP-generators, electronic noise generators, etc.

Shnoll (2009, 2012) concludes that factors determining shapes of modeled histograms may facilitate to reveal factors that determine the shapes of physical histograms; the physical processes, to which the changes of histogram shapes are due, in turn can be attributed to the motion of the Earth through cosmic space.

Figure 7 depicts fluctuations of the decay constant of Tritium caused by non-stationarity cosmic factors of unkmown origin (recorded by our team), which is complete in agreement with observations of Shnoll (2009, 2012).

One more study of cosmic factors is related to the pioneering works of Kozyrev (Kozyrev 1977; Kozyrev and Nasonov 1978) who demonstrated the existence of a remote influence of a star, which affected the measuring transmitter (a resister), much early than the star became visible in the telescope. Those first results of 1970s then were confirmed by other researchers (Lavrentiev, Yeganova et al. 1990; Lavrentiev, Gusev et al. 1990) who noted that Kozyrev's and their observations of planets, stars and galaxies showed that in fact there exists a kind of an interaction, which modern physics does not consider. By their opinion the investigation of this kind of the interaction has an important meaning for the development of ideas of physics about the real constitution of space-time.

So, what was the reason for those Kozyrev's enigmatic superluminal rays? It seems plausible that they are related to free inertons irradiated by stars, i.e. inertons that escaping from the appropriate inerton clouds of stars and planets at fast non-adiabatic processes.

Free inertons liberated from inerton clouds of particles can migrate through space until other massive objects will absorb them. Since inertons are carriers of mass properties of matter, it will be interesting to launch astrophysical observations using appropriate inerton antennas, receivers, electronics and software. Inerton observatories will be able to yield pictures of the universe derived from inerton radiation (instead of the existing laboratories constructed for the detection of hypothetical gravitational waves, incomprehensible dark matter particles and other exotic particles).

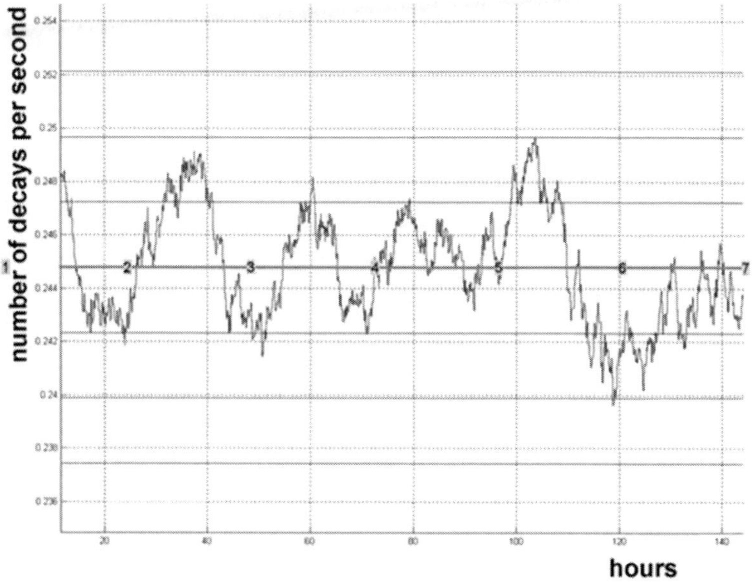

Figure 7. Fluctuations of the decay constant of Tritium recorded by our research team in 2010. The days of measurements are applied on the horizontal axis.

Figure 8 demonstrates the device 'Rudra' designed by our research team, which measures inertons. In the device two kinds of the antenna are used: a ferrite rod-type core and a disc-type piezoceramic. The measurement depends on the orientation of the antenna. The antenna absorbs a signal and the density of the antenna's material increases locally for a short time. Mechanical changes are transformed in the appropriate magnetic/electric polarization, which immediately influences the current of an electronic scheme. An amplifier increases the current to the level sufficient to launch the channel of signal recording. The frequency divider, interchanging switch and timer perform the adapting between the level of the signal and the discharge grid of the recorder. The device is designed on the principle of a counter: it counts arriving inertons that strike the antenna. When the antenna is put is a metal box that screens it of electromagnetic waves, the device continues to operate counting inerton signals. However, if a standard radio used for normal radio reception is placed in the same metal box, the receiver stops to function altogether.

Figure 9 demonstrates our recent device that can be called an inerton gravity-gradient meter. Its antenna consists of two quartz crystals functioning at the frequency of 10 MHz with an accuracy of 10^{-12}. Each of the crystals is set in its proper metal casing that completely screens the detector from the electromagnetic environment. A change in the orientation of one of the detectors immediately shows a disharmony between the resonance frequencies of 10 MHz. Since the device measures only inertons, this disharmony correlates with the density of an inerton pulse or inerton flow caught by the detector.

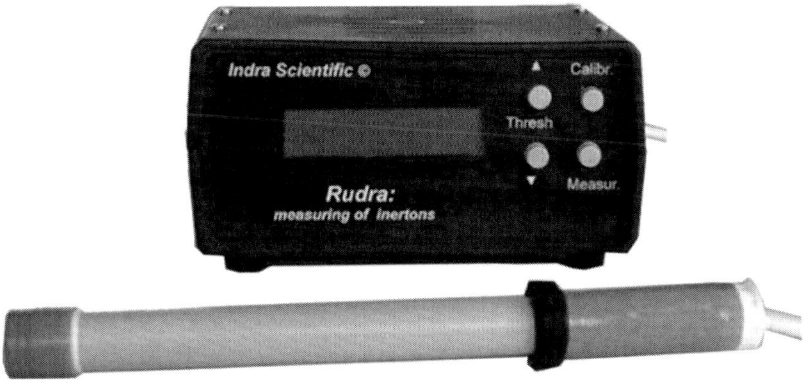

Figure 8. The ‘Rudra’ inerton measuring device. The device is able to count inerton signals received by the antenna per second (i.e. can measure the intensity) and can also show the spectrum of absorbed inerton signals from a few Hertz to 100 kHz.

Below, Figs 10-13, we present results of our measurements as a function of the antenna orientation and the time of measurement. The measurements were performed with apredecessor of the Rudra device. The measurements were on purpose conducted at difficult conditions in a windowless room in a concrete building. Obvious measurements of the Earth’s inerton field exhibit the pronounced orientation effect: signals picked up along the West-East line are more intensive than signals received along the North-South line. A gradual decrease in the intensity of inertons along the West-East line and the increase of the intensity along the North-South line with time is associated with the proper rotation of the Earth when the antenna orientation with respect to the Sun gradually changes. In case of the West-East line of the antenna orientation the dip angle of inertons coming from the Sun begins to depart from normal, which means the decrease in absorption of inertons. In case of the North-South line of the antenna orientation, the dip angle of inertons coming from the Sun tends to the normal giving rise to an additional absorption of inertons.

With the measurement of terrestrial inertons we can trace periodical modulations caused by the influence of solar inerton rays, Figs. 10 to 14. Fluctuations may be associated with the influence of solar inerton activity. Solar inertons introduce appropriate changes to the intensity of terrestrial inertons.

These changes vary from 5 to 20 % probably owing to the rotation of the Earth. The most significant rise in inerton intensity is observed during the light period of the astronomical day. Meteorological factors also severely affect the measurements, which means that clouds saturated with water, the motion of clouds strengthened by wind and the rain are additional sources of inerton radiation on the surface of the Earth.

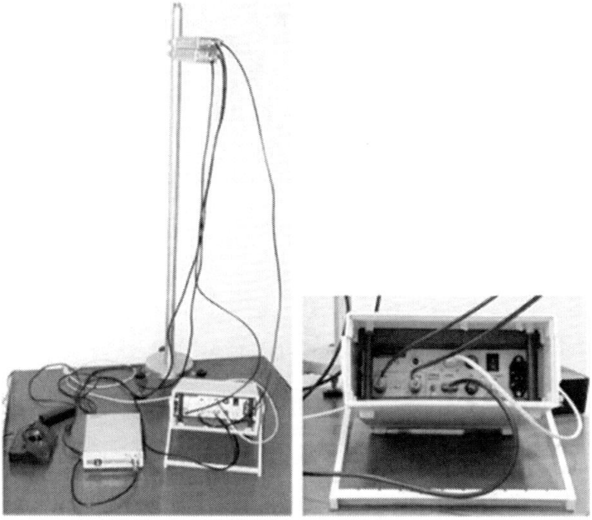

Figure 9. Inerton gravity-gradient meter.

Figures 14 and 15 exhibit the results of measurements obtained with our modern version of the 'Rudra' device (Figure 8) at two mutually perpendicular positions of the antenna: the North-South line and the West-East line. The measurements were carried out in a cabin in a village south of Kyiv, Ukraine (latitude: 50 North, longitude: 30.5 East). There were no electromagnetic power sources in a radius of 25 km. The measurements were carried out on 26 November 2011.

Figure 14 depicts data recorded early morning in the dark circa one hour before dawn. At the orientation of the antenna's surface along the West-East line the device recorded 82 ± 10 impulses per second with a ferrite antenna with a coil and 90 ± 10 impulses per second for the piezoceramic sensor. At the orientation of the two antennas along the North-South line, there were recorded 76 ± 10 and 69 ± 10 impulses per second, respectively.

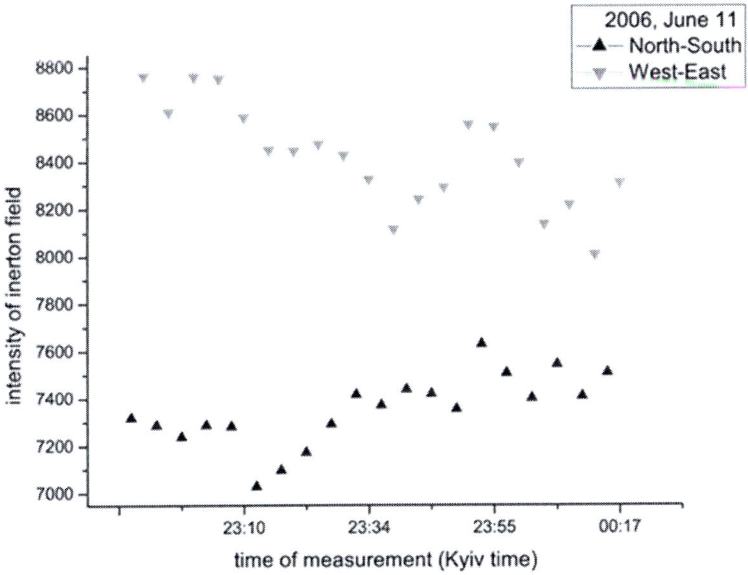

Figure 10. Intensity of inerton flow versus time of measurement. Quite warm night.

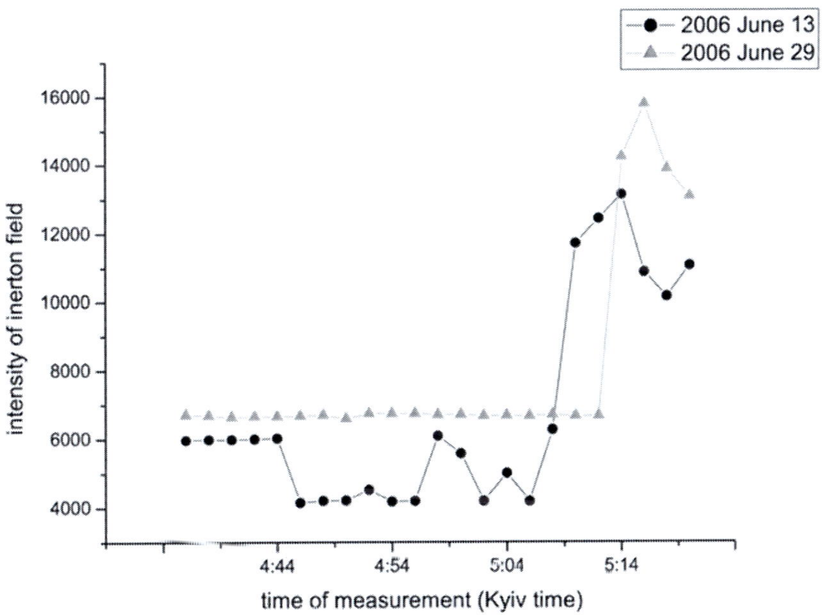

Figure 11. Intensity of the inerton field along the West-East line versus time. Nice and sunny days. Sharp splashes correspond to the time of sunrise.

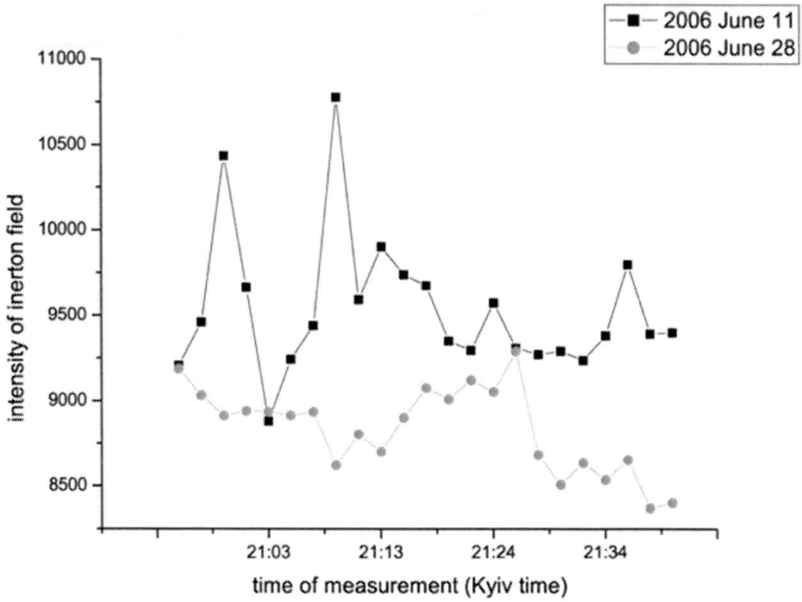

Figure 12. Intensity of inerton field at the moment of sunset versus time.

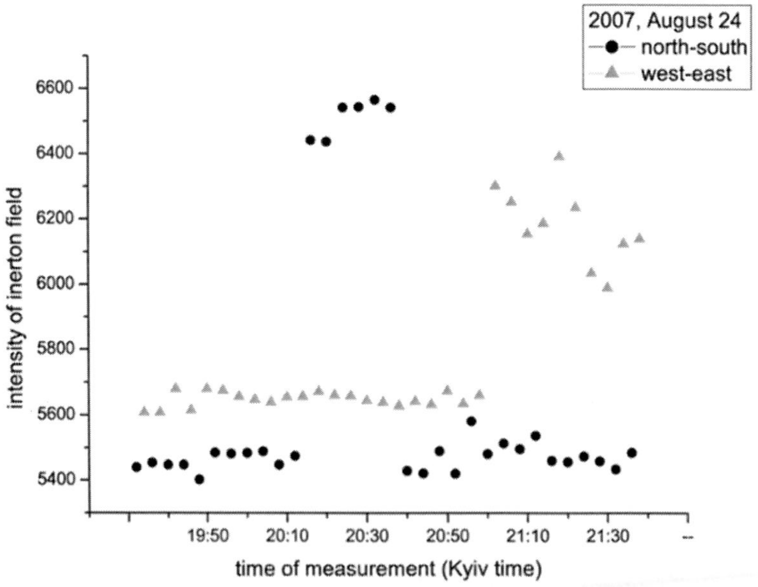

Figure 13. Intensity of inerton field at the moment of sunset versus time.

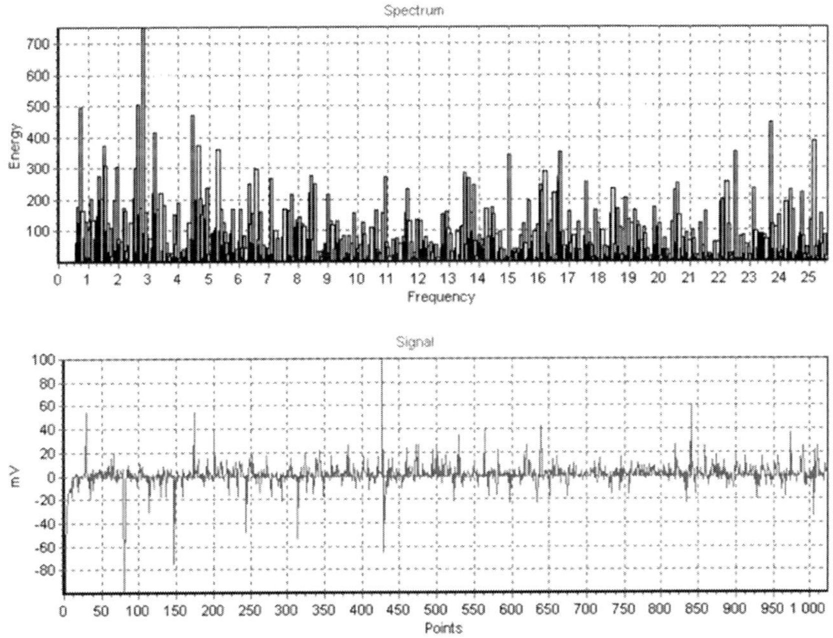

Figure 14. Record of inerton signals (bottom) and its frequency spectrum decoding (in kHz, top) in the dark at 5:57 a.m. The antenna is the piezoceramic disc; the orientation "West-East".

One hour later (Figure 15) in the dawn time we could see that the intensity of inertons significantly increased, as the energy of inerton impulses reached 5000 a.u. When the antennas were aligned along the West-East line the intensity of the inerton field increased by 400% (340 ± 30 impulses per second), while with the orientation along the North-South line it showed an increase 82% (92 ± 10 impulses per second).

Basing on our measurements, we may conclude that the spectral decomposition of the recorded signals in the range of a few Hz to 100 kHz exhibits the densest inerton zone between 2 and 4 kHz. When the orientation of the antennas coincides with the West-East direction, the mentioned zone of frequencies is more clearly expressed than at the orientation of the antennas along the North-South line.

The 'Rudra' device can further be redesigned for astronomical observations. In particular, the behavior of inerton activity of the Sun looks extremely interesting and important; because the Sun is not only a globe of plasma and inner nuclear reactions, it is also the source of a huge dynamic mass, 1.99 x 10^{30} kg. This mass irradiates flows of inertons of different intensity and frequencies. We measured an influence of the solar inerton radiation on the Earth's background inerton field. We constantly observed changes in the Earth inerton filed associated with instability of atmospheric processes. For example, we recorded an unusual change in the spectrum associated with the Lunar eclipse that occurred in our place (Kyiv, Ukraine) on 15 June 2011.

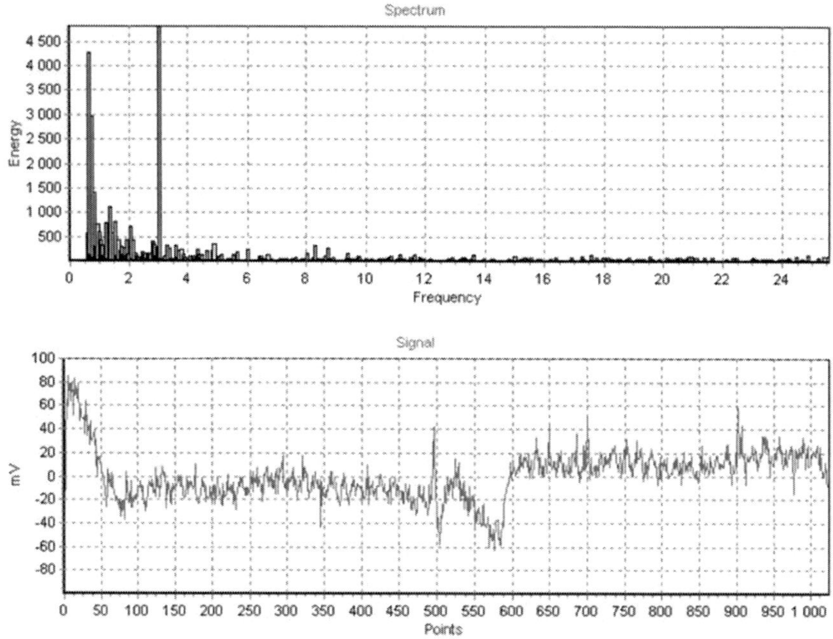

Figure 15. Record of inerton signals (bottom) and its frequency spectrum decoding (in kHz, top) during dawn at 6:56 a.m. The antenna is the ferrite rod with a coil; the orientation “West-East”.

Thus, inertons from outer space can be recorded at a laboratory on the Earth. We may study the intensity of inerton signals, the direction from which they arrive, their frequency, the degree of homogeneity of signals and we may also measure the velocity of inertons, which by preliminary estimation is about two orders higher than the speed of light *c*. In our approach we are able to carry out a multidimensional analysis by using a one-dimensional channel (i.e. one procedure of measurement). Subsequently we can extract even very weak signals (with signal to noise ratio of up to 1:1000 using special technology). We measure amplitude, frequency spectrum and time characteristics.

We are interested to collaborate with astronomers and astrophysicists in designing the first measuring system, an inerton observatory, which will be able to show amplitude, spectral and time characteristics of non-stationary signals in real time. The system will record inerton signals from the Earth atmosphere, from the Earth, the Sun as well as distant objects in the celestial sphere by means of different methods, namely:

- measurement of anomalies by using detectors on the basis of a piezoceramic disc or and/or a ferrite rod with a copper coil in a range of Hz up to hundreds of MHz;
- measurement using a high precision inerton gravity-gradient meter made on the basis of quartz crystals, developed and designed by our team (it can distinguish signals from background noise with a sensitivity of up to 10^{-12}); the vertical gravity gradient data can be gauged with an accuracy better than 0.1 Eötvös;
- measurement of cosmic anomalies by means of the point-by-point gauging of background radiation with a sensitivity of 10^{-10} using photomultipliers(radiation range from a few keV to many GeV);
- measurement using of Terra-hertz receivers.

In conditions of fixed or variable geometry we have to distinguish an anomaly that by its characteristics will be significantly weaker than the general background (i.e. we have to separate a very weak signal from the noise of the Earth, a star, nebula, etc.). We are able to do this with one-time measurement. The noise can be significantly higher than the signal that we will be searching for. The direction of the signal can also be undetermined, which requires special conditions: the necessity to have a position-sensitive component. Thus, the system will consist of a tracking facility and the four abovementioned measuring set-ups.

Conclusion

So far the universe has not been considered as a substrate that is governed by strict mathematical laws. The present work shows that real space can be described in the form of the tessel-lattice of primary elements, topological balls. Matter in the universe is a deformation of the tessel-lattice and hence any interaction between material objects involves space.

Dark energy is nothing but a peculiar distribution of topological balls in our universe considered as a huge-scale cluster in the total tessel-lattice.

The gravitational interaction, as a basic interaction in the universe, is derived directly from the behavior of the object studied in the tessel-lattice. The gravitational interaction between many objects induces in addition to the corrected Newton potential (38) also an elastic interaction. These two factors are responsible for the phenomenon known as dark matter, namely, a distribution of standing inerton waves around massive bodies creates an obstacle for travelling photons.

Until today the universe has been explored only in different spectra of photon radiation. Now a new possibility is presented – an examination of outer space through inerton rays, which are mass carriers of the gravitational interaction of cosmic objects. The investigation will require new kinds of astronomical equipment and appropriate inerton observatories, gateways to future inerton astronomy.

Acknowledgments

The author greatly thanks colleagues Yu. Zabulonov and V. Burtniak for the assistance in designing the devices for measuring inerton fields. Many thanks to A. Keder for critical reading of the manuscript.

References

Abazajian K. N., et al. (2012). *Light sterile neutrinos: A white paper*, arXiv:1204.5379 [hep-ph].

Albrecht A., Bernstein G., Cahn R., Freedman W. L., Hewitt J., Hu W., Huth J., Kamionkowski M., Kolb E. W., Knox L., Mather J. C., Staggs S. and Suntzeff N. B. (2006). *Report of the dark energy task force*, arXiv: astro-ph/0609591.

Anderson J. D., Laing P. A., Lau E. L., Liu A. S., Nieto M. M., and Turyshev S. G. (2002). Study of the anomalous acceleration of Pioneer 10 and 11, *Phys. Rev. D* 65, 082004 (55 p.); arXiv: gr-qc/0104064.

Andreev E., Dovbeshko G. and Krasnoholovets V. (2007). The study of influence of the Teslar technology on aqueous solution of some biomolecules, *Research Lett. Phys. Chem.*, Volume 2007, Article ID 94286, 5 pages; arXiv:1204.6062 [physics.gen-ph].

Angus G. W. and Diaferio A. (2011). The abundance of galaxy clusters in MOND: Cosmological simulations with massive neutrinos, *Mon. Not. R. Astron. Soc.*417, no. 2, 941-949; arXiv:1104.5040 [astro-ph.CO].

Aprile E., et al. (The XENON100 Collaboration) (2011). Implications on inelastic dark matter from 100 live days of XENON100 data. *Phys. Rev. D*84 (061101).

Baretti Machín R. (2012). http://www.uprh.edu/rbaretti/Nordstrom3dic2012.htm.

Barrow J. D. (2003). Unusual features of varying speed of light cosmologies, *Phys. Lett.* B 564, 1-7; arXiv: gr-qc/0211074.

Bernabei R., Belli P., Cappella F., et al. (2010). New results from DAMA/LIBRA, *Eur. Phys. J. C* 67, 39–49.

Bernabei R., Belli P., Cappella F., Caracciolo V., Cerulli R., Dai C. J., d'Angelo A., Di Marco A., He H. L., Incicchitti A., Ma X. H., Montecchia F., Sheng X. D., Wang R. G., Ye Z. P. (2012). No role for muons in the DAMA annual modulation results, *Eur. Phys. J. C* 72, 2064.

Benítez-Llambay A., Navarro J. F., Abadi M. G., Gottlöber S., Yepes G., Hoffman Y., and Steinmetz M. (2013). Dwarf galaxies and the cosmic web, *Astr. J. Lett.* 763 (2), L41.

Bounias M. and Krasnoholovets V. (2003a). Scanning the structure of ill-known spaces: Part 1. Founding principles about mathematical constitution of space, *Kybernetes: The Int. J. Systems & Cybernetics* 32, no. 7/8, 945-975, Eds.: L. Feng, B. P. Gibson and Yi Lin; also arXiv: physics/0211096.

Bounias M. and Krasnoholovets V. (2003b). Scanning the structure of ill-known spaces: Part 2. Principles of construction of physical space, *Kybernetes: The Int. J. Systems & Cybernetics* 32, no. 7/8, 976-1004, Eds.: L. Feng, B. P. Gibson and Yi Lin; also arXiv:physics/0212004.

Bounias M. and Krasnoholovets V. (2003c). Scanning the structure of ill-known spaces: Part 3. Distribution of topological structures at elementary and cosmic scales, *Kybernetes: The Int. J. Systems & Cybernetics* 32, no. 7/8, 1005-1020, Eds.: L. Feng, B. P. Gibson and Yi Lin; also arXiv: physics/0301049.

Bounias M. and Krasnoholovets V. (2004). The universe from nothing: A mathematical lattice of empty sets, *Int. J. Anticipatory Computing Systems* 16, 3-24, Ed.: D. Dubois; arXiv:physics/0309102.

Boyarsky A., Ruchayskiy O. and Shaposhnikov M. (2009). The Role of Sterile Neutrinos in Cosmology and Astrophysics, *Ann. Rev. Nuclear and Particle Science* 59: 191-214; arXiv:0901.0011 [hep-ph].

Boylan-Kolchin M., Besla G., HernquistL. (2011). Dynamics of the Magellanic Clouds in a Lambda cold dark matter universe, *Month. Not. Royal Astr. Soc.* 414, no. 2, 1560–1572.

Brooks A. M.Kuhlen M., Zolotov A., Hooper D. (2013). A baryonic solution to the missing satellites problem, *Astrophys. J.* 765, 22; arXiv:1209.5394 [astro-phys.CO].

CDMS Collaboration. (2011).Results from a low-energy analysis of the CDMS II Germanium data. arXiv:1011.2482 [astro-phys.CO].

Casado J. (2003). A simple cosmological model with decreasing light speed, arXiv:astro-ph/0310178 [astro-ph].

Clowe D., Gonzalez A. and Markevich M. (2004). Weak-lensing mass reconstruction of the interacting cluster 1E0657-558: Direct evidence for the existence of dark matter, *Astrophys. J.* 604, 596-603.

COBE satellite. (2009). http://lambda.gsfc.nasa.gov/product/cobe/

Crothers S. J. (2005). On the ramifications of the Schwarzschild space-time metric, *Progress in Physics* 1, 74-80, http://www.ptep-online.com/index_files/complete/PP-01-2005.pdf.

Crothers S. J. (2010).The black hole, the Big Bang: A cosmology in crisis, Proc. NPA, http://vixra.org/pdf/1103.0047v1.pdf

Crothers S. J. (2011). The black hole catastrophe: A short reply to J. J. Sharples, *Hadronic J.* 34, 197-224 (2011); also http://vixra.org/pdf/1206.0080v2.pdf.

Crothers S. J. (2012a). Proof of no "black hole" binary in Nova Scorpii, *Global Journal of Science Frontier Research Physics and Space Science* Volume 12, no. 4 Version 1.0 June 2012; http://vixra.org/pdf/1206.0080v2.pdf.

Crothers S. J. (2012b). General relativity – a theory in crisis, *Global Journal of Science Frontier Research Physics and Space Science* Volume 12 Issue 4 Version 1.0 June 2012, http://vixra.org/pdf/1207.0018v2.pdf.

Crothers S. J. (2012c). *Proof of the invalidity of the black hole and Einstein's field equations* http://vixra.org/pdf/1212.0060v1.pdf.

Guy B. (2010). A modified law of gravitation taking account of the relative speeds of moving masses. *A preliminary study*, http://hal.archives-ouvertes.fr/hal-00472210.

de Broglie L. (1925). Recherches sur la théorie des quanta, *Ann. de Phys.*, 10e série, t. III, 22-128 (Janvie-Février); translation by A. F. Kracklauer: *On the theory of quanta*, (Lulu.com; Morrisville, NC; 2007).

de Broglie L. (1967). Sur la Dynamique du corps a masse propre variable et la formule de transformation relativiste de la chaleaur, *Comptes Rendus* 264 B, no. 15, 1173-1175.

de Broglie L. (1986). *Heisenberg's uncertainty relations and the probabilistic interpretation of wave mechanics*, Mir, Moscow, pp. 42-43 (Russian translation).

de Vega H. J., Sánchez N. G. (2011). Warm dark matter in the galaxies: Theoretical and observational progresses. *Highlights and conclusions of the Chalonge CIAS Meudon Workshop* 2011 Ecole Internationale d'Astrophysique Daniel Chalonge Meudon campus of Observatoire de Paris in the historic Château, 8-10 June 2011, arXiv:1109.3187 [astro-phys.CO].

de Vega H. J., Moreno O., Moya de Guerra E., Medrano, M. R., and Sánchez N. G. (2013). Role of sterile neutrino warm dark matter in Rhenium and Tritium beta decays, *Nucl. Phys. B* 866, 177-195.

Einasto J. (2010). Dark matter, in *Astronomy and Astrophysics*, Eds.: O. Engvold, J. Lattanzio, B. Czerny and R. Stabell, The EOLSS On-line is a virtual library http://www.eolss.net/outlinecomponents/Astronomy-Astrophysics.aspx, arXiv:0901.0632 [astro-ph.CO].

Einstein A. (1916). Die Grundlage der allgemeinen Relativitätstheorie, *Annalen der Physik* 354, no. 7, 769-822.

Ellis G. and Uzan J. (2003). c is the speed of light, isn't it?, *Am. J. Phys.* 73, 240-247; arXiv: gr-qc/0305099.

Famaey B. and McGaugh S. (2012). Modified Newtonian dynamics (MOND): Observational phenomenology and relativistic extensions, arXiv:1112.3960 [astro-ph.CO].

Famaey B. and McGaugh S. (2013). *Challenges for ΛCDM and MOND*, arXiv:1301.0623 [astro-ph.CO].

Feng J. I. and Kumar J. (2008). The WIMPless miracle: Dark matter particles without weak-scale masses or weak interactions, *Phys. Rev. Lett.* 101, 231301, 4 pages; arXiv:0803.4196 [hep-ph].

Feng J. Q., Gallo C.F. (2008). *Galactic rotation described with various thin-disk gravitational models*. arXiv:0804.0217 [astro-ph].

Feng J. Q., Gallo C.F. (2011). Modeling the Newtonian dynamics for rotation curve analysis of thin-disk galaxies, *Res. Astronomy and Astrophysics*, 11, no. 12, 1429-1448.

Fort B.and Mellier Y. (1994). Arc(let)s in clusters of galaxies, *Astronomy and Astrophys. Rev.* 5, 239-292.

Freedman W. L.(2000). Determination of cosmological parameters, *Phys. Scripta* 85, 37-46; arXiv:astro-ph/9905222.

Friedmann A. (1922). Über die Krümmung des Raumes, *Zeitschrift für Physik* A10 (1): 377–386.

Garret K. and Dūda G. (2011). Dark matter: A primer, *Advances in Astronomy*, Volume 2011, Article ID 968283, 22 pages.

Ghosh P. P., Mukhopadhyay U. and Ray S. (2012). Does accelerating Universe permit varying speed of light? *Astrophys. Space Sci.*337, no. 2, 509-510.

Gilmore G., Wilkinson M. I., Wyse R. F. G., Kleyna J. T., Koch A., Evans N. W., Grebel E. K., (2007).The observed properties of dark matter on small spatial scales, *Astrophys. J.* 663, 948-959.

Hajdukovic, D. S. (2011). Is dark matter an illusion created by the gravitational polarization of the quantum vacuum?, *Astrophys. Space Sci.* 334, 215-218; arXiv:1106.0847 [physics.gen-ph].

Hofer W. A. (2012). Heisenberg, uncertainty, and the scanning tunneling microscope, *Frontiers of Physics* 7, 218-222; arXiv:1105.3914.

Hoyle, F.; Burbidge, G.; Narlikar, J. V. (1993). A quasi-steady state cosmological model with creation of matter, *Astrophys. J.* 410, 437–457.

Hoyle, F.; Burbidge, G.; Narlikar, J. V. (1994). Further astrophysical quantities expected in a quasi-steady state Universe, *Astronomy Astrophys.*289, no. 3, 729–739.

Hubble E. (1929). A relation between distance and radial velocity among extra-galactic nebula, *Proc. Nat. Acad. Scien.* 15, no. 3, 168-173.

Jee M. J., Ford H. C., Illingworth G. D., White R. L., Broadhurst T. J., Coe D. A., Meurer G. R., van der Wela, Benítez N., Blakeslee J. P., Bouwens R. J., Bradley L. D., Demarco R., Homeier N. L., Martel A. R., and Mei S. (2007). Discovery of a ringlike dark matter structure in the core of the galaxy cluster Cl 0024+17, *Astrophys. J.* 661, 728-749.

Kamada A., Yoshida N., Kohri K., Takahashi T. (2013). Structure of dark matter halos in warm dark matter models and in models with long-lived charged massive particles, arXiv:1301.2744 [astro-phys.CO].

Kozyrev N. A. (1977). Astronomical observations by means of the physical properties of time, in: *Flaring Stars: Proceedings of the Symposium dedicated to the openeing2.6-m telescope of the Byurakan Astrophysical Observatory*, Byurakan, October 5-8, 1976. Yerevan, pp. 209-227; in Russian.

Kozyrev N. A. and Nasonov V. V. (1978). A new method of determination of trigonometric parallaxes on the basis of measurements of difference between apparent and visible position of stars, in: *Astronometry and Celestial Mechanics*, Akademiya Nauk SSSR, Moscow, Leningrad, pp. 168-179.

Krasinsky G. A. and Brumberg G.A. (2004). Secular increase of astronomical unit from analysis of the major planets motions, and its interpretation. *Celest. Mech. & Dyn. Astron.*, 90, 267.

Krasnoholovets V. (1997). Motion of a relativistic particle and the vacuum, *Phys. Essays* 10, no. 3, 407-416; arXiv:quant-ph/9903077.

Krasnoholovets V. (2001a). On the theory of the anomalous photoelectric effect stemming from a substructure of matter waves, *Ind. J. Theor. Phys.*49, no. 1, 1-32; arXiv:quant-ph/9906091.

Krasnoholovets V. (2001b). On the mass of elementary carriers of gravitational interaction, *Spacetime & Substance* 2, no. 4, 169-170; arXiv:quant-ph/0201131.

Krasnoholovets V. (2002).Submicroscopic deterministic quantum mechanics, *Int. J. Computing Anticipatory Systems* 11, 164-179; arXiv:quant-ph/0109012.

Krasnoholovets V. (2003). On the nature of the electric charge, *Hadronic J. Suppl.* 18, no. 4, 425-456; arXiv.org:physics/0501132.

Krasnoholovets V. (2008). Reasons for the gravitational mass and the problem of quantum gravity. In *Ether space-time and cosmology*, Vol. 1; Duffy M., Levy J. and Krasnoholovets V., Eds.; PD Publications: Liverpool, pp. 419-450; arXiv:1104.5270 [physics.gen-ph].

Krasnoholovets V. (2009). On microscopic interpretation of phenomena predicted by the formalism of general relativity, in *Ether space-time and cosmology*, Vol. 2: *New insights into a key physical medium.* Eds.: Duffy M. C. and Lévy J. (Apeiron, 2009), pp. 417-431 [Totally 488 pages. Publisher: C. Roy Keys Inc. (Apeiron); also in *Apeiron* 16, no. 3, 418-438 http://redshift.vif.com/JournalFiles/V16NO3PDF/V16N3KRA.pdf.

Krasnoholovets V. (2010). Inerton fields: Very new ideas on fundamental physics, *American Inst. Phys. Conf. Proc.* - December 22, Vol. 1316, 244-268. Search for fundamental theory: The VII international symposium honoring French mathematical physicist Jean-Pierre Vigier (12-14 July 2010, Imperial College, London), Eds.: R. L. Amoroso, P. Rowlands and S. Jeffers.

Krasnoholovets V. (2011). Dark matter as seen from the physical point of view, *Astrophys. Space Sci.* 335, no. 2, 619-627.

Krasnoholovets V. (2013a). Inerton field effects in nanosystems, in *Research Horizons of Nanosystems: Structure, Properties and Interactions*, Ed.: M. Putz, Apple Academy Press and CRC Press, Taylor & Francis Group, in press.

Krasnoholovets V. (2013b). On the gravitational time delay effect and the curvature of space, *Int. J. Computing Anticipatory Systems*, in press.

Krasnoholovets V. and Byckov V. (2000). Real inertons against hypothetical gravitons. Experimental proof of the existence of inertons, *Ind. J. Theor. Phys.*48, no. 1, 1-23; arXiv.org:quant-ph/0007027.

Krasnoholovets V. and Ivanovsky D. (1993). Motion of a particle and the vacuum, *Phys. Essays* 6, no. 4, 554-563; arXiv:quant-ph/9910023.

Krasnoholovets V., Skliarenko S. and Strokach O. (2006). On the behavior of physical parameters of aqueous solutions affected by the inerton field of Teslar technology, *Int. J. Mod. Phys. B* 20, no. 1, 111-124; arxiv:0810.2005 [physics.gen-ph].

Krasnoholovets V., Kukhtarev N. and Kukhtareva T. (2006). Heavy electrons: Electron droplets generated by photogalvanic and pyroelectric effects, *Int. J. Mod. Phys. B* 20, no. 16, 2323-2337; arXiv:0911.2361 [quant-ph].

Krasnoholovets V. and Lev B. (2003). Systems of particles with interaction and the cluster formation in condensed matter, *Cond. Matt. Phys.* 6, 67-83; arXiv:cond-mat/0210131.

Kroupa, P. (2012). The dark matter crisis: Falsification of the current standard model of cosmology. *Publications of the Astronomical Society of Australia* 29, no. 4, 395-433; arXiv:1204.2546 [astro-ph.CO].

Kroupa P., Pawlowski M., Milgrom M. (2012) The failures of the standard model of cosmology require a new paradigm. *Int. J. Mod. Phys. D* 21, 1230003 [13 pages]; arXiv:1301.3907 [astro-ph.CO].

Lavrentiev M. M., Yeganova I. E., Lutset M. K. and Fominykh S. F. (1990). About distant influence of stars on a resistor. *Proceed. Acad. Scien. USSR* 314, no. 2, 352-355; in Russian.

Lavrentiev M. M., Gusev V. A., Yeganova I. A., Lutset M. K. and Fominykh S. F. (1990). About the registration of an actual position of the Sun, *Proc. Acad. Sci. USSR* 315, no. 2, 368–370; in Russian.

Lämmerzahl C., Preuss O., and Dittus H. (2008). *Is the physics within the Solar system really understood?* arXiv:gr-qc/0604052.

Lemaître G. (1931). Expansion of the universe. A homogeneous universe of constant mass and increasing radius accounting for the radial velocity of extra-galactic nebulæ, *Mon. Not. R. Astron. Soc.*91, 483–490.

Loinger A. (2002). *On black holes and gravitational waves* (La Goliardica Pavese).

Loinger A. (2007). *More on BH's and GW's*.III (La Goliardica Pavese).

Magueijo J. (2003). New varying speed of light theories, *Rep. Progr. Phys.* 66, 2025-2068 (2003), arXiv:astro-ph/0305457.

Mannheim P. D. and O'Brien J. G. (2010). *Fitting galactic rotation curves with conformal gravity and a global quadratic potential*, Arxiv:1011.3495 [astro-ph.CO].

Mannheim P. D. and O'Brien J. G. (2011). Impact of a global quadratic potential on galactic rotation curves, *Phys. Rev. Lett.* 106, no. 12, 121101.

McCulloch M. E. (2012). Testing quantised inertia on galactic scales, arXiv:1207.7007 [physics.gen-ph].

Merritt D. and Bertone, G. (2005). Dark matter dynamics and indirect detection. *Mod. Phys. Lett.* A20, no. 14, 1021–1036.

Milgrom M. (1983). A modi_cation of the Newtonian dynamics as a possible alternative to the hidden mass hypothesis, *Astrophys. J.* 270, 365-370.

Milgrom A. (2009). The MOND limit from space-time scale invariance, *Astrophys. J.* 698, 1630-1639; arXiv: 0810.4065 [astro-phys.].

Nesbet R. K. (2013). Conformal gravity: Dark matter and dark energy, *Entropy* 15, 162-176.

Penzias A. A. and Wilson R. W. (1965). A measurement of excess antenna temperature at 4080 Mc/s, *Astrophys. J.* 142, 419–421.

Perlmutter S., Aldering G., Goldhaber G., Knop R. A., Nugent P., Castro P. G., Deustua S., Fabbro S., Goobar A., Groom D. E., Hook I. M., Kim A. G., Kim M. Y., Lee J. C., Nunes N. J., Pain R., Pennypacker C. R., Quimby R., Lidman C., Ellis R. S., Irwin M., McMahon R. G., Ruiz-Lapuente P., Walton N., Schaefer B., Boyle B. J., Filippenko A. V., Matheson T., Fruchter A. S., Panagia N., Newberg H. J. M., Couch W. J. (The Supernova Cosmology Project). (1999). Measurements of Ω and Λ from 42 high-redshift supernovae, *Astrophys. J.* 517, no. 2, 565-586; arXiv:astro-ph/9812133.

Petit J. P. and d'Agostini G. (2007). *Bigravity: a bimetric model of the Universe with variable constants, including VSL (variable speed of light)*, arXiv:0803.1362 [physics.gen-ph].

Poncaré H. (1906) Sur la dynamique de l'électron, *Rendiconti del Circolo matematico di Palermo* 21, 129-176; also: Oeuvres, t. IX, pp. 494-550 (in Russian translation: *Selected Transactions*, ed. N. N. Bogolubov (Nauka, Moscow, 1974), 3, 429-486).

Rabounski D. (2008). On the current situation concerning the black hole problem. *Progress in Physics* 1, 101–103.

Robertson H. P. (1935). Kinematics and world structure, *Astrophys. J.*82, 284–301.

Robitaille P.-M. (2007). WMAP: A radiological analysis, *Progress in Physics* 1, no. 1, 3-18.

Robitaille P.-M. (2009). COBE: A radiological analysis, *Progress in Physics* 4, no. 1, 17-42.

Robitaille P.-M. (2010a). Calibration of microwave reference blackbodies and targets for use in satellite observations: An analysis of errors in theoretical outlooks and testing procedures *Progress in Physics,* 3, no. 1, 3-10.

Robitaille P.-M. (2010b). The Planck satellite LFI and the microwave background: importance of the 4K reference targets, *Progress in Physics* 3, no. 1, 11-18.

Rocha G., Contaldi C. R., Colombo L. P. L., Bond J. R., Gorski K. M., and Lawrence C. R. (2010). Application of XFASTER power spectrum and likelihood estimator to Planck, *Mon. Not. R. Astron. Soc.* 414, 823-846; arXiv:1008.4948 [astro-ph.CO].

Rocha M., Peter A. H. G., Bullock J. S., Kaplinghat M., Garrison-Kimmel S., Oorbe J., and Moustakas L. A. (2012). Cosmological simulations with self-interacting dark matter I: Constant density cores and substructure, arXiv:1208.3025 [astro-ph.CO].

Salih M. A., Tomsah I. B. I., Al Mahdi T. M. (2012). General relativity in microscopic scale (GRMS), *J. Basic. Appl. Sci. Res.* 2, no. 12, 12348-12353.

Santilli R. M. (2006). *Isodual theory of antimatter with applications to antigravity, grand unification, and cosmology*, Kluwer Academic Publishers, Boston, Dordrecht, London, pp. 25-28, 187.

Schmidt B. P., Suntzeff N. B., Phillips M. M., Schommer R. A., Clocchiatti A., Kirshner R. P., Garnavich P., Challis P., Leibundgut B., Spyromilio J., Riess A. G., Filippenko A. V., Hamuy M., Smith R. C., Hogan C., Stubbs C., Diercks A., Reiss D., Gilliland R., Tonry J., Maza J., Dressler A., Walsh J., Ciardullo R.(1998).The high-z supernova search: Measuring cosmic deceleration and global curvature of the universe using type IA supernovae, *Astrophys. J.* 507, no. 1, 46-63; arXiv:astro-ph/9805200.

Schwarzschild K. (1916). Uber das Gravitationsfeld eines Massenpunktes nach der Einsteinschen Theorie. *Sitzungsberichte der Kniglich Preussischen Akademie derWissenschaften*1, 189-196.

Shnol S. E. (2001). Macroscopic fluctuations are possible consequences of fluctuations of space-time. Arithmetical and cosmological physical aspects, *Rossiyskiy Khimicheski Zhurnal* 65, no. 1, 12-15.

Shnoll S. E. (2009). *Cosmophysical factors in random processes*, Svenska fysikarkivet, Täby (in Russian). Second edition, supplemented: (2012) American Research Press, Rehoboth, New Mexico, USA.

Shnoll S. E., Rubinshtejn I. A., Zenchenko K. I., Shlekhtarev V. A, Kaminsky A. V., Konradov A. A., and Udaltsova N. V. (2005). Experiments with rotating collimators cutting out pencil of α-particles at radioactive decay of 239Pu evidence sharp anisotropy of space, *Progr. Phys.* vol. 1, 81–84; arXiv:physics/0501004 [physics.space-ph].

Shnol S. E., Zenchenko T. A., Zenchenko K. I., Pozharski E. V., Kolombet V. A. and Konradov A. A. (2000). Regular changes offine structure of statistical distributions as a consequence of cosmological physical factors, *Uspekhi Fiz. Nauk* 170, no. 2, 214-218.

Stasielak J., Biermann P. L., and Kusenko A.(2006*). Thermal evolution of the primordial clouds in warm dark matter models with keV sterile neutrinos*, arXiv: astro-ph/0606435.

The Top 30 Problems with the Big Bang. (2002). *Apeiron* 9, no. 2, 72-90. http://redshift.vif.com/ JournalFiles/V09NO2PDF/V09N2tvf.PDF.

Thornhill C. K. (2001). A non-singular ethereal cosmology, *Hadronic J. Suppl.*16, 203-262.

Tinker J., Kravtsov A. V., Klypin A., Abazajian K., Warren M., Yepes G., Gottlöber S., Holz D. E. (2008). Toward a halo mass function for precision cosmology: the limits of universality, *Astrophys. J.* 688, 709-728.

Toffolatti L., Burigana C., Argüeso F., and Diego J. M. (2012). Extragalactic compact sources in the Planck sky and their cosmological implications, in *Open Questions in Cosmology*, G. J. Olmo (Ed.), (IFIC - Spain) and published by InTechOpen, Chapter 3, p. 57-86 (2012); arXiv:1302.3355 [astro-phys.CO].

Turyshev S. G., Toth V. T., Kinsella G., LeeS.-C., Lok S. M. and Ellis J. (2012). Support for the thermal origin of the Pioneer anomaly, *Phys. Rev. Lett.* 108, 241101 (5 pages); arXiv:1204.2507 [gr-qc].

Turyshev S. G., Toth V. T. and Sazhin M. V. (2013). General relativistic observables of the GRAIL mission, *Phys. Rev.* D 87 024020 (29 p.); arXiv:1212.0232 [gr-qc].

Van Dyck R. S., Jr., Ekstrom P., and Dehmelt H G. (1976). Axial, magnetron, cyclotron and spincyclotron beat frequencies measured on single electron almost at rest in free space (Geonium), *Nature* 262, 776–777.

Van Dyck R. S., Jr., Schwinberg P. B. and Dehmelt H. G. (1986). Electron magnetic moment from geonium spectra: Early experiments and background concepts. *Phys. Rev. D* 34, 722–736.

Volovik, G. (2003). *The universe in a Helium droplet*, Clarendon Press, Oxford.

Walker, A. G. (1937). On Milne's theory of world-structure, *Proc. London Math. Soc.*s2-42, no. 1, 90–127.

Wilczek F. (2009). What is space? *Physics @MIT* 30, http://web.mit.edu/ physics/people/faculty/docs/ wilczek_space06.pdf.

WMAP satellite. Tests of Big Bang: The CMB. (2011). http://map.gsfc.nasa.gov/ universe/bb_tests_cmb.html.

Zolotov A., Brooks A. M., Willman B., Governato F., Pontzen A., Christensen C., Dekel A., Quinn T., Shen S., and Wadsley J. (2012). *Astrophys. J.* 761, 71; arXiv:1207.0007 [astro-phys.CO].

In: Research Development in Dark Matter Research ISBN: 978-1-62948-010-7
Editors: Nori Kinjo and Akira Nakajima

Chapter 2

SOME NEW ASTRONOMICAL CONSEQUENCES OF THE DARK MATTER CONCEPT

L. M. Chechin
V G. Fessenkov Astrophysical Institute,
National Centre for Space Researches
and Technologies of the National Space Agency,
Almaty, Kazakhstan

Abstract

Aims: This chapter is dedicated to completely considering the new model of the halo theoretical description of dark matter, including new consequences of dark matter influence on the electromagnetic wave propagation – light rays and Cerenkov radiation.

In its introduction we have also briefly considered a few other astronomical consequences of the existence of dark matter as discussed in modern literature.

Study design: Astronomical consequences of the dark matter concept.

Place and Duration of Study: V. G. Fessenkov Astrophysical Institute, during 2012 year.

Methodology: proposing the simplest form of central symmetric potential celestial – mechanical type, i.e. $U(\varphi) = -\mu / \varphi$; clarifying that positions of images of lensing objects depend on profiles of dark matter and baryonic matter; searching the Cherenkov radiation in a galaxy's halo of dark matter, described by profiles of Navarro-Frenk-White and Burkert.

Results: we argued that the standard field scalar potential couldn't be widely used for getting adequate curve lines of the galaxy and determining the profiles of dark matter in their halo. To discover the global properties of scalar fields that can describe the observable characteristics of dark matter on cosmological space and time scales, we proposed the simplest form of central symmetric potential celestial – mechanical type, i.e. $U(\varphi) = -\mu / \varphi$. It was shown that this potential allows us to attain dark matter profiles and rotational curve lines for dwarf galaxies. We also point out that these are in good agreement with some previous results, based on the N-body simulation method.

The light rays' propagation is searched in the spherical-symmetric gravitational field of the Galaxy described both by dark matter as well as baryonic matter. We focused on the case when contributions on the gravitational lensing of these components are comparable to each other. It is shown that positions of images of lensing objects depend on profiles of dark matter and baryonic matter. It was also argued that positions of images from the gravitational lens are drastically smaller than the distance between the gravitational lens and observer.

The Cherenkov radiation in a galaxy's halo of dark matter, described by profiles of Navarro-Frenk-White and Burkert was also searched. Powers of the Cherenkov radiation for the refractive indexes of gravitational fields with these profiles were calculated. It was shown that correspondent radiation temperatures in the X-rays diapason have the magnitude of about $T \sim 10^{-6} K$. It was also marked that its detection may be the criteria for choosing the preferable dark matter density distribution in a galaxy.

Conclusion: some new astronomical consequences of the dark matter concept have been considered in detail. It was shown that our observable predictions lay in the threshold of modern astronomical measurements.

Keywords: Scalar field; dark matter profile; halo of baryonic mass profile; gravitational lens; the Cherenkov radiation

1. Dark-Matter Halo as the Gravitational Lens Model

Introduction

According to modern concepts, the general morphological structure of a galaxy has the following components: a central part (galaxy core), a bulge surrounding it, a gas-dust disk, a star cluster, and a dark-matter halo [1]. It is important to emphasize that the dark-matter halo comprises the main part of the galaxy mass (up to 90%).

In this subsection the light propagation in a galaxy is investigated taking into account that the character of light propagation is influenced only by the dark matter. This means that light propagates far from the central part of the galaxy whose sizes are designated by *R*. We assume that the size of the galaxy as a whole equals r_0. Then the region of light rays propagation is determined by the conditions $r_0 >> r >> R$. From this it follows that $\frac{r}{r_0} << 1$.

To investigate the deviation of light rays in the dark-matter halo we need to know its spatial distribution.

In the literature, a number of dark-matter profiles are well known. Among them are the Navarro–Frenk–White profile[2], the Burkert profile [3], the Kravtsov–Klypin profile [4], and others [5, 6]. It is quite clear that each of these profile cause different effects of light ray propagation. This relates to the fact that the galaxy, due to its mass, leads to curvature of the space-time surrounding it. And the curved space-time, according to [7], can be considered as an original medium with the corresponding effective refractive index.

The purpose of this subsection is investigation of light rays propagating in the dark-matter halos described using different profiles and comparing the refractive indexes to find the largest deviation angle of the light rays. In its turn, this allows us to refine the theory of gravitational microlensing.

1.1. Spherically Symmetric Dark-Matter Halos as a Gravitational Lens Model

The gravitational lens refers to any distribution of matter – stars, galaxies, and especially clusters of galaxies, with their essentially more massive dark matter halos – between a distant source and an observer capable of bending the light from a source, as it travels towards the observer. Space-time around a massive object is curved, and as a result light rays from a background source propagating through space-time are bent. The simplest gravitational lens is the spherically symmetric one.

Let write down the general form of the metric of spherically symmetric gravitation field

$$dS^2 = -e^{\lambda(r)} dr^2 - r^2 (d\theta^2 + \sin^2\theta d\varphi^2) + e^{\nu(r)} c^2 dt^2 \qquad (1.1)$$

where, according to [8],

$$e^{\lambda(r)} = 1 + \frac{8\pi G}{c^2 r} \int_0^r \rho(r) r^2 dr \ , \qquad (1.2)$$

$$e^{\nu(r)} = \exp\int_r^\infty \left(\frac{8\pi G}{c^2} (\rho(r) + p) r e^{\lambda} - \frac{d\lambda}{dr} \right) dr . \tag{1.3}$$

Here $\rho(r)$ and $p(r)$ are the density of matter and its pressure, respectively. To solve our problem, it is necessary to assign the concrete expressions for these quantities. In introduction it already has been noted that a number of dark-matter profiles are known in the literature. To achieve the objective of the present study, we take advantage of only some of them.

Concerning the pressure concretization we set $p(r) = 0$ that relates to the model of dark matter as an ideal fluid.

i) For the Navarro–Frenk–White profile [2]

$$\rho(r) = \frac{\rho_0}{\frac{r}{r_0}\left(1 + \frac{r}{r_0}\right)^2} , \tag{1.4}$$

we have the following solution

$$e^{\lambda(r)} = 1 + \frac{8\pi G}{c^2 r} \int_0^r \frac{\rho_0}{\frac{r}{r_0}\left(1 + \frac{r}{r_0}\right)^2} r^2 dr , \tag{1.5}$$

$$e^{\nu(r)} = \exp\int_r^\infty \left(\frac{8\pi G}{c^2} \frac{\rho_0}{\frac{r}{r_0}\left(1 + \frac{r}{r_0}\right)^2} r e^{\lambda} - \frac{d\lambda}{dr} \right) dr . \tag{1.6}$$

Here in after, ρ_0 is the dark-matter density in the galaxy's center. To calculate these integrals, we put the condition $\frac{r}{r_0} << 1$ which allows expansion of the integrand in the Taylor series. Below we restrict ourselves by consideration of terms whose order is not higher than $\frac{8\pi G}{c^2}\rho_0 r_0^2\left(\frac{r}{r_0}\right)$. With this accuracy we have

$$e^{\lambda(r)} = 1 + \frac{4\pi G}{c^2}\rho_0 r_0^2\left(\frac{r}{r_0}\right), \tag{1.7}$$

$$e^{\nu(r)} = 1 + \frac{8\pi G}{c^2}\rho_0 r_0^2 - \frac{4\pi G}{c^2}\rho_0 r_0^2\left(\frac{r}{r_0}\right). \tag{1.8}$$

Then the metric of the centrally symmetric gravitation field of the dark-matter halo is represented in the form

$$\begin{aligned} dS^2 &= -\left[1 + \frac{4\pi G}{c^2}\rho_0 r_0 r\right]dr^2 - r^2(d\theta^2 + \sin^2\theta\, d\varphi^2) \\ &+ \left[1 + \frac{8\pi G}{c^2}\rho_0 r_0^2 - \frac{4\pi G}{c^2}\rho_0 r_0 r\right]c^2 dt^2 \end{aligned} \quad . \tag{1.9}$$

Recall that in orderto investigate the light rays propagation in a certain metric – to determine the refractive index of the gravitation field – it is necessary to equate its 4-dimensional interval to zero[7]. Since the refractive index in the centrally symmetric metric depends only on the radius, assume that $\theta = \frac{\pi}{2}$ and $\varphi = 0$. Then from equation (1.9) we have

$$-\left[1 + \frac{4\pi G}{c^2}\rho_0 r_0 r\right] dr^2 + \left[1 + \frac{8\pi G}{c^2}\rho_0 r_0^2 - \frac{4\pi G}{c^2}\rho_0 r_0 r\right]c^2 dt^2 = 0\,. \tag{1.10}$$

Introducing the velocity of light in the medium

$$\upsilon = \frac{dr}{dt}, \tag{1.11}$$

from equation (1.10) we obtain

$$\left[1+\frac{8\pi G}{c^2}\rho_0 r_0^2-\frac{4\pi G}{c^2}\rho_0 r_0 r\right]c^2=\left[1+\frac{4\pi G}{c^2}\rho_0 r_0 r\right]\upsilon^2 . \tag{1.12}$$

Basing on the refractive index definition

$$n=\frac{c}{\upsilon}, \tag{1.13}$$

from equation (1.12) with the above-indicated accuracy we find its value

$$n=1+\frac{4\pi G}{c^2}\rho_0 r_0^2\left(1+\frac{r}{r_0}\right). \tag{1.14}$$

From this it follows that the refractive index, in addition to its main – constant – component $\left(\frac{4\pi G}{c^2}\rho_0 r_0^2\right)$ comprises the term directly proportional to the distance.

ii) We now consider the Burkert profile

$$\rho(r)=\frac{\rho_0}{\left(1+\frac{r}{r_0}\right)\left(1+\frac{r^2}{r_0^2}\right)}, \tag{1.15}$$

so that

$$e^{-\lambda(r)}=1-\frac{8\pi G}{c^2 r}\int_0^r\frac{\rho_0}{\left(1+\frac{r}{r_0}\right)\left(1+\frac{r^2}{r_0^2}\right)}r^2 dr , \tag{1.16}$$

$$e^{\nu(r)} = \exp\int_{r}^{\infty}\left(\frac{8\pi G}{c^2}\frac{\rho_0}{\left(1+\frac{r}{r_0}\right)\left(1+\frac{r^2}{r_0^2}\right)}re^{\lambda} - \frac{d\lambda}{dr}\right)dr\,. \qquad (1.17)$$

To calculate these integrals, we also set the condition $\frac{r}{r_0} << 1$ that allows us to expand the integrand in the Taylor series and consider terms of an order not greater than the $\frac{8\pi G}{c^2}\rho_0 r_0^2\left(\frac{r}{r_0}\right)$. With this accuracy, we have

$$e^{\lambda(r)} = 1, \qquad (1.18)$$

$$e^{\nu(r)} = 1 + \frac{8\pi G}{c^2}\rho_0 r_0^2\left(\frac{r}{r_0}\right). \qquad (1.19)$$

Hence the metric of a centrally symmetric gravitation field of dark-matter's halo has the following form

$$dS^2 = -dr^2 - r^2(d\theta^2 + \sin^2\theta\ d\varphi^2) + \left[1 + \frac{8\pi G}{c^2}\rho_0 r_0 r\right] c^2 dt^2 \qquad (1.20)$$

Again we take advantage of the condition that for investigating the light ray propagation in a certain metric – for determining the refractive index of a gravitationfield – it is necessary to equate its 4-dimensional interval to zero [7]. Assuming that $\theta = \frac{\pi}{2}$ and $\varphi = 0$, from expression (1.20) we have

$$dr^2 = \left[1 + \frac{8\pi G}{c^2}\rho_0 r_0^2\left(\frac{r}{r_0}\right)\right]c^2 dt^2\,. \qquad (1.21)$$

Using expression (1.11) for the velocity of light in medium, from equation (1.21) we obtain

$$\left[1+\frac{8\pi G}{c^2}\rho_0 r_0^2\left(\frac{r}{r_0}\right)\right] c^2=\upsilon^2 . \tag{1.22}$$

Analogous to the previous case, by introducing the refractive index of medium $n=\frac{c}{\upsilon}$, from equation (22) with the above-indicated accuracy, we find its value

$$n=1+\frac{4\pi G}{c^2}\rho_0 r_0^2\left(\frac{r}{r_0}\right). \tag{1.23}$$

From here it follows that the refractive index is still directly proportional to the distance.

iii) Finally, we consider the Kravtsov–Klypin profile [5]

$$\rho(r)=\frac{\rho_0}{\left(\frac{r}{r_0}\right)^{\gamma}\left[1+\left(\frac{r}{r_0}\right)^{\alpha}\right]^{\frac{\beta-\gamma}{\alpha}}}. \tag{1.24}$$

Here, unlike the Navarro–Frenk–White and Burkert profiles, three unknown coefficients - α, β and γ are already used. This uncertainty allows investigation of more general types of dark-matter profiles. To find the metric of the gravitation field generated by the Kravtsov–Klypin distribution, take the Newton binomial

$$(a+b)^n=a^n+na^{n-1}b+\frac{n(n-1)}{2}a^{n-2}b^2+...+b^n\,.. \tag{1.25}$$

Using it, we expand the expression in the denominator enclosed in the square brackets of (1.24). Also for accounting the condition $\frac{r}{r_0}\ll 1$ that is used everywhere, set $\alpha=1$. Then we obtain

$$\rho(r)=\rho_0\left[\frac{1}{\left(\frac{r}{r_0}\right)^{\gamma}}-(\beta-\gamma)(\beta-\gamma-1)\left(\frac{r}{r_0}\right)^{\beta-2\gamma-2}\right]. \tag{1.26}$$

Substitution of this expression into equation (1.2) and (1.3), leads to the following coefficients of metric tensor

$$e^{-\lambda(r)}=1-\frac{8\pi G}{c^2 r}\int_0^r\left[\frac{\rho_0}{\left(\frac{r}{r_0}\right)^{\gamma}}-(\beta-\gamma)(\beta-\gamma-1)\rho_0\left(\frac{r}{r_0}\right)^{\beta-2\gamma-2}\right]r^2dr\,, \tag{1.27}$$

$$e^{\nu(r)}=\exp\int_r^{\infty}\left[\frac{8\pi G}{c^2}\left(\frac{\rho_0}{\left(\frac{r}{r_0}\right)^{\gamma}}-(\beta-\gamma)(\beta-\gamma-1)\rho_0\left(\frac{r}{r_0}\right)^{\beta-2\gamma-2}re^{\lambda}\right)-\frac{d\lambda}{dr}\right]dr\,. \tag{1.28}$$

Performing all necessary calculations with the required accuracy we obtain the explicit form of these coefficients

$$e^{\lambda(r)}=1+\frac{1}{3-\gamma}\frac{8\pi G}{c^2}\rho_0 r_0^2\left[\left(\frac{r}{r_0}\right)^{2-\gamma}-\frac{(\beta-\gamma)(\beta-\gamma-1)}{(\beta-2\gamma+1)}\left(\frac{r}{r_0}\right)^{\beta-2\gamma}\right], \tag{1.29}$$

$$e^{\nu(r)}=1+\frac{8\pi G}{c^2}\rho_0 r_0^2\frac{1}{2-\gamma}\left(1-\frac{5-2\gamma}{3-\gamma}\left(\frac{r}{r_0}\right)^{2-\gamma}\right)-$$
$$-\frac{8\pi G}{c^2}\rho_0 r_0^2\frac{(\beta-\gamma)(\beta-\gamma-1)}{(\beta-2\gamma)}\left(1-\frac{2\beta-4\beta+1}{\beta-2\gamma+1}\left(\frac{r}{r_0}\right)^{\beta-2\gamma}\right). \tag{1.30}$$

Thus, the refractive index for the Kravtsov–Klypin profile is

$$n(r)=1+\frac{4\pi G}{c^2}\rho_0 r_0^2\frac{(\beta-\gamma-1)}{(\beta-2\gamma)}\left(1-\frac{3\beta-6\beta+1}{\beta-2\gamma+1}\left(\frac{r}{r_0}\right)^{\beta-2\gamma}\right)-$$
$$-\frac{4\pi G}{c^2}\rho_0 r_0^2\frac{1}{(2-\gamma)}\left(1-\frac{7-3\beta}{3-\gamma}\left(\frac{r}{r_0}\right)^{2-\gamma}\right) \quad . \qquad (1.31)$$

1.2. Deviation of Light Rays in Galaxy Dark-Matter Halos

To find the deviation of light rays in the gravitation fields of galaxy dark-matter halos that were obtained above, let's use the general expression from optics of inhomogeneous medium [9]

$$\Delta\theta=\int_{-\infty}^{\infty}\frac{d}{dr}(\ln n)dr=2\ln n(r)\Big|_{-\infty}^{r}. \qquad (1.32)$$

Substituting here, one by one, all refractive indices founded above, we obtain

$$\Delta\theta_{NFW}=2\frac{4\pi G}{c^2}\rho_0 r_0^2\left(1+\frac{r}{r_0}\right), \qquad (1.33)$$

$$\Delta\theta_{B}=2\frac{4\pi G}{c^2}\rho_0 r_0^2\left(\frac{r}{r_0}\right), \qquad (1.34)$$

$$\Delta\theta_{KK}=2\frac{4\pi G}{c^2}\rho_0 r_0^2\frac{(\beta-\gamma)(\beta-\gamma+1)}{\beta-2\gamma}\left(1-\frac{3\beta-6\gamma+1}{\beta-2\gamma+1}\left(\frac{r}{r_0}\right)^{\beta-2\gamma}\right)-$$
$$-2\frac{4\pi G}{c^2}\rho_0 r_0^2\frac{1}{(2-\gamma)}\left(1-\frac{7-3\gamma}{3-\gamma}\left(\frac{r}{r_0}\right)^{2-\gamma}\right) \qquad (1.35)$$

To estimate values of these angles let us assume that the average dark-matter density is $\rho_0\sim10^{-23}g/cm^3$ [6], the typical galaxy size is $r_0\sim50Mpc$, and its current radius is $r \sim 0.5pc$. Remembering that $\frac{G}{c^2} \approx 10^{-28} cm/g$ and $\frac{r}{r_0} \approx 10^{-2}$, we obtain the estimation for the main factor - $\frac{4\pi G}{c^2}\rho_0 r_0^2 \approx 3 \cdot 10^{-5}$. For the Navarro–Frenk–White and Burkert deviation angles we obtain the following estimations - $\Delta\theta_{NFW} \approx 6 \cdot 10^{-5}$ and $\Delta\theta_B \approx 6 \cdot 10^{-5}$, accordingly. In angular units $(1'' \approx 5 \cdot 10^{-6})$ these deviation angles are

$$\Delta\theta_{NFW} \approx +12'', \tag{1.36}$$

$$\Delta\theta_B \approx +12''. \tag{1.37}$$

As for the Kravtsov–Klypin deviation angle that describes by expression (1.35) with a set of unknown coefficients, we have

$$\Delta\theta_{KK} = +12''\left(\frac{(\beta-\gamma)(\beta-\gamma+1)}{\beta-2\gamma} - \frac{1}{2-\gamma}\right). \tag{1.38}$$

Conclusion

Based on the obtained results, we can conclude the following.

Firstly, the deviation angles seem to be very large. Indeed, the deviation angle of light rays in the Sun's gravitation field is $\Delta\theta = 1.75''$ [8] and is approximately 10 times less than the angles given by equation (1.36) and (1.37). However, our numerical estimates essentially depend on the dark-matter density. If we use its value $\rho_{DM} \sim 10^{-30} g/cm^3$ averaged over the entire Universe [1], the deviation angles become smaller and in radians will be in the order of $\sim 10^{-12}$ or in angular units $\sim 0.00001''$. We note that the accuracy of gravitational lensing of galaxies reached at the present time is about $0.0001''$.

Secondly, for the Navarro–Frenk–White, Burkert and Kravtsov–Klypin profiles the light rays are focused, because $\Delta\theta > 0$. Moreover, it is clear that expression (1.38) basically depends on the values of coefficients β and γthat shows the essential indefinity of deflecting angle's magnitude. For these reasons it is impossible to give the likely estimation for it.

2. On the Theory of Two-component Gravitational Lens

Introduction

Previously, it was considered that the basic components of a galaxy are the central part (nucleus of a galaxy), bulge, gaseous-dust disc, stars cluster and halo of dark matter [10]. Remember that dark matter is about ninety percent of the total mass of a galaxy. That is why the influence of dark matter on the gravitational lensing effect was the topic of many researchers [11, 12].

But if light rays are propagating in the vicinity of the central part of a galaxy we must take into account the influence of baryonic matter on the gravitational lensing effect. We will identify such lens as a two-component gravitational lens.

To examining this lens it is necessary know the distribution of baryonic matter in a galaxy. Though this problem was researched by a number of authors, we limit ourselves to the King [13] and Idliss [14] distributions. As for the profile for dark matter we'll take the well-known Navarro-Frenk- White profile, only [2].

The focus of the present subchapter is searching the propagation of the light rays in the spherical-symmetric gravitational field of the galaxy described by both dark matter and baryonic matter. We are examining the case when contributions on the gravitational lensing of these components are comparable to each other. Later on the movement of light rays will be considered in the coordinate plane $\theta=\frac{\pi}{2}$, for simplicity.

The basic equation that describes light rays propagation throughout the gravitational lens is the aberrational equation. It connects the impact parameter p with the deflecting angle θ and distance P from the observer and light rays' trajectory. All of the subsequent denotations are corresponding to the denotations in monograph [11].

The general form of aberrational equation is the algebraic equation of $n-$ th order. To compose the aberrational equation, note that accordingly [11] the following expression takes place

$$P_i = p + \Delta P_i, \qquad (i=1,2) \tag{2.1}$$

where $\Delta P_i -$ is the light rays displacement from the rectilinear trajectory in the gravitational lens. Due to the minuteness of the deflecting angle we have the approximate relation

$$\Delta P = x \cdot tg\theta \approx x \cdot \theta \,, \tag{2.2}$$

where $x -$ is the position of the observer behind the gravitational lens on its principal axis.

For the following analyses note that deflecting angle θ must depend on the impact parameter P, i.e. $\theta = \theta(P)$. In fact, according to general relativity the deflecting angle in the gravitational field of an opaque body (Sun, for example) is inversely proportional to the impact parameter, i.e. $\theta(P) = \frac{\alpha}{P}$, where $\alpha = \frac{4\pi GM}{c^2}$. In our article [15] it was shown that the deflecting angle might depend on the impact parameter in the first order as well as in the second order. That is why for the transparent cosmic object (galaxy) we set the parametric representation of deflecting angle

$$\theta(P) = \alpha_0 P^0 + \alpha_1 P^1 + \alpha_2 P^2 + ... + \alpha_n P^n = \alpha_0 + \alpha_1 P + \alpha_2 P^2 + ... + \alpha_n P^n \,. \tag{2.3}$$

Substituting (2.3) into (2.1) and (2.2) anyone can get the general expression for the aberrational equation

$$P = p + (\alpha_0 + \alpha_1 P^1 + \alpha_2 P^2 + ... + \alpha_n P^n)x \,. \tag{2.4}$$

The choice of aberrational equation order depends on the numerical magnitudes of coefficients α_0, α_1, α_2, ..., α_n. In our article we'll consider the aberrational equation of order not higher than two.

So, let the aberrational equation of the second order

$$\alpha_2 x P^2 + (\alpha_1 x - 1)P + \alpha_0 x = 0 \tag{2.5}$$

symbolized in the standard reduced form

$$P^2+\left(\frac{\alpha_1-\frac{1}{x}}{\alpha_2}\right)p+\frac{\alpha_0}{\alpha_2}=0\cdot \tag{2.6}$$

Its two roots are ($x=R_0$)

$$\left.\begin{aligned} P_1&=\frac{1}{2}\left(\frac{\alpha-\frac{1}{R_0}}{\alpha_2}\right)-\sqrt{\frac{1}{4}\left(\frac{\alpha-\frac{1}{R_0}}{\alpha_2}\right)^2-\frac{\alpha_0}{\alpha_2}}\\ P_2&=\frac{1}{2}\left(\frac{\alpha-\frac{1}{R_0}}{\alpha_2}\right)+\sqrt{\frac{1}{4}\left(\frac{\alpha-\frac{1}{R_0}}{\alpha_2}\right)^2-\frac{\alpha_0}{\alpha_2}}\end{aligned}\right\}. \tag{2.7}$$

And at last, for the linear aberrational equation

$$P=\alpha_0\cdot x+\alpha_1 P\cdot x \tag{2.8}$$

the impact parameter determines by expression $\left(x=R_0\right)$

$$P=-\frac{\alpha_0 R_0}{\alpha_1 R_0-1}=-\frac{\alpha_0}{\alpha_1-\frac{1}{R}}\cdot \tag{2.9}$$

Based on these results it is possible to calculate the actual parameters $\alpha_0,\alpha_1,\alpha_2$ for the profiles mentioned before.

Using the multiplicative representation of the refractive index, argued in [11], for the two-component substance (dark matter and baryonic matter) we write down the approximate expression for the general refractive index as

$$n=n_{DM}n_{BM}\approx 1+\eta_{DM}+\eta_{BM}\ , \tag{2.10}$$

where

$$\left.\begin{aligned} n_{DM} &= 1 + \eta_{DM} \\ n_{BM} &- 1 + \eta_{BM} \end{aligned}\right\}. \tag{2.11}$$

The mentioned profiles combine into two groups – NFW-profile and Idliss' distribution, NFW-profile and King's distribution.

From [15] it follows that the refractive index of the galaxy with NFW-profile for dark matter and baryonic matter distribution according to Idliss has the form

$$n_{NFW-Id} = 1 + \frac{8\pi G}{c^2}\rho_0 r_0^2 - \frac{8\pi G}{5c^2}\rho_0 r_0^2\left(\frac{r}{r_0}\right) - \frac{4\pi G}{3c^2}\rho_0^{'} r_0^2\left(\frac{r^2}{r_0^2}\right). \tag{2.12}$$

Tp estimate the baryonic matter density in the center of the galaxy let's take the results of article [16]. As mass of the central part of our Galaxy is about $3 \cdot 10^5 M_{\odot}$ in the cubic parsec, then the central baryonic mass density will be $\rho'_0 = 3 \cdot 10^5 M_{\odot} / 4 \cdot 10^{66} cm^3 \approx 1 \cdot 10^{-28} g / cm^3$.

Using the well-known optics [9] expression for the deflection angle $\Delta\theta = \int_{-\infty}^{\infty} \frac{d}{dr}(\ln n)dr = 2 \ln n\Big|_{-\infty}^{r}$, we find the total angle

$$\theta_{NFW-Id} = 2\left(\frac{8\pi G}{c^2}\rho_0 r_0^2 - \frac{8\pi G}{5c^2}\rho_0 r_0^2\left(\frac{r}{r_0}\right) - \frac{4\pi G}{3c^2}\rho'_0\, r_0^2\left(\frac{r^2}{r_0^2}\right)\right). \tag{2.13}$$

In this expression the numerical magnitudes of coefficients are –

$$\left.\begin{aligned} \alpha_0 &= \frac{8\pi G}{c^2}\rho_0 r_0^2 = 16\pi \\ \alpha_1 &= -\frac{8\pi G}{5c^2}\rho_0 r_0 = -3\pi \cdot 10^{-23}\left[cm^{-1}\right] \\ \alpha_2 &= -\frac{4\pi G}{3c^2}\rho_0{}' = -3\pi \cdot 10^{-46}\left[cm^{-2}\right] \end{aligned}\right\} \tag{2.14}$$

Putting them into (2.13) we get the angle's magnitude

$$\theta_{NFW-Id} \cong 4'',3 . \tag{2.15}$$

Set the test position of the observer is described by magnitude $x = 3.7 \cdot 10^{27}\,cm$ (position of quasar QSO 0957+561). As parameters that include into (7) have the following numerical magnitudes

$$\frac{1}{2}\left(\frac{\alpha_1 - \frac{1}{x}}{\alpha_2}\right) = \frac{1}{2} \times 10^{23}\,cm, \tag{2.16}$$

$$\frac{\alpha_0}{\alpha_2} = -5.3 \times 10^{46}\,cm^2 , \tag{2.17}$$

the roots of aberrational equation (or impact parameters), hence, are

$$\left.\begin{array}{l} P_1 \approx 2.8 \times 10^{23}\,cm, \\ P_2 \approx -1.8 \times 10^{23}\,cm \end{array}\right\}. \tag{2.18}$$

From these calculations anyone can see that the impact parameter (position of a star image) has a positive magnitude as well as a negative magnitude. In doing this, their modules are not equal to each other. Therefore the observer will see two images of one and the same object that are located on opposite sides of the gravitational lens' principal optical axis. Note that they are located at distances that are drastically smaller than the test position of the observer - $P_1 \cong 2.8 \cdot 10^{23}\,cm \ll x = 3.7 \cdot 10^{27}\,cm$ and $P_2 \cong -1.8 \cdot 10^{23}\,cm \ll x = 3.7 \cdot 10^{27}\,cm$.

Further, the refractive index of a galaxy with NFW–profile of dark matter and King's distribution of baryonic matter has the form

$$n_{NFW-K} = 1 + \frac{8\pi G}{c^2}\rho_0 r_0^2 - \frac{8\pi G}{3c^2}\rho''_0\, r_0^2 - \frac{8\pi G}{5c^2}\rho r_0^2\left(\frac{r}{r_0}\right) + \frac{8\pi G}{c^2}\rho''_0\, r_0^2\left(\frac{r}{r_0}\right). \tag{2.19}$$

The corresponding deflection angle will be

$$\theta_{NFW-K} = \left(\frac{8\pi G}{c^2} \rho_0 r_0^2 - \frac{8\pi G}{3c^2} \rho''_0 r_0^2 - \frac{8\pi G}{5c^2} \rho r_0^2 \left(\frac{r}{r_0} \right) + \frac{8\pi G}{c^2} \rho''_0 r_0^2 \left(\frac{r}{r_0} \right) \right). \quad (2.20)$$

(Here $\rho_0'' \sim \rho_0'$ is the central baryonic mass density of the King distribution). Using the previous results we find the desired coefficients in the expression (20)

$$\left. \begin{aligned} \alpha_{0_1} &= \frac{8\pi G}{c^2} \rho_0 r_0^2 \approx 16\pi \\ \alpha_{0_2} &= \frac{8\pi G}{3c^2} \rho_0'' r_0^2 \approx -5\pi \\ \alpha_{1_1} &= \frac{8\pi G}{5c^2} \rho_0 r_0^2 \approx -3.2\pi \cdot 10^{-23} \left[cм^{-1} \right] \\ \alpha_{1_2} &= \frac{8\pi G}{c^2} \rho_0'' r_0^2 \approx 16\pi \cdot 10^{-23} \left[cм^{-1} \right] \end{aligned} \right\} \quad (2.21)$$

and the numerical magnitude of the deflection angle

$$\theta_{NFW-K} \cong 2'',6 . \quad (2.22)$$

That is why the roots of the aberrational equation will be

$$P_1 = -\frac{\alpha_{0_1}}{\alpha_{1_1} - \dfrac{1}{R_0}} \approx 5.3 \cdot 10^{23} cm , \quad (2.23)$$

$$P_2 = -\frac{\alpha_{0_1}}{\alpha_{1_1} - \dfrac{1}{R_0}} \approx 0.3 \cdot 10^{23} cm . \quad (2.24)$$

From (2.23) and (2.24) we will see that the lensing star has two images that are close each other and lay on the same side of the principal optical axis. But as in the previous case they are placed at distances that are essentially smaller then the test position of the observer - $P_1 \approx 5.3 \cdot 10^{23} cm \ll x = 3.7 \cdot 10^{27} cm$ and $P_1 \approx 0.3 \cdot 10^{23} cm \ll x = 3.7 \cdot 10^{27} cm$.

Conclusion

In conclusion, the main results of this subchapter are: first – the positions of images of lensing objects depend on profiles of dark matter as well as baryonic matter. Second – the positions of images of lensing objects are drastically smaller than the distance of the observer from the gravitational lens.

3. The Cherenkov Radiation in the Galaxy's Halo of Dark Matter

3.1. On the Theory of Cherenkov Radiation in a Gravitational Field

The well-known substance-refracting index is the value that is equal to the ratio between phase velocities in a vacuum and in a given medium. However, if any charge particle whose group velocity v in a medium is larger than the phase velocity in this substance c, i.e. $v > c$, the specific electromagnetic radiation appears. The theory of such radiation – the Cherenkov radiation – was given in the references [17, 18], for example.

The generalization of Cherenkov radiation on a gravitational field presence was given in monograph [19], where it was shown that for the Schwarzschild field this radiation is characterized by the upper boundary frequency $\omega = pc\sqrt{1 - \frac{2m}{r}} \frac{r}{m}\left(1 - \frac{c}{v}\sqrt{1 - \frac{2m}{r}}\right)$. Note that this effect is very small and may be detected in the very strong gravitational field. (It is necessary to cite the article [20] where it was shown that if a charged particle will move in the external gravitational field along the geodesic line it creates the Cherenkov radiation also, except for the case of the field's nonzero Ricci tensor.)

By the way, the problems of searching the behavior of electromagnetic radiation and light rays, not only in the local weak gravitational fields but also on the cosmological scale where they were essentially developed last time. it is enough to mention the search for the gravitational micro lensing of the stars' halo of dark matter. In spite of the minuteness of the angle images' separation (about 0,"001) it is possible to detect the stars' brightness changing under the mutual movement of star, galaxy and observer at the micro lensing process (see, for example [21]).

Another aspect of dark matter searching on cosmological scales, based on the different radiation processes studying in it, include the dynamics of cosmic rays in dark matter, as in articles [22]. Moreover, the search for properties of dark matter on the small space scales are seen in articles [23, 24].

Next the dark matter density in the vicinity of the Sun was discussed in [25], where the influence of dark matter on the light rays' propagation in the Solar system was also considered. In fact, in this article the time delay effect (analog of the Shapiro effect) and the shift of light rays' frequency effect in the spherical-symmetric dark matter distribution have been considered. But the shape of dark matter profile was chosen a priori as $\rho(t,r)=\rho(t)\left(\frac{\ell}{r}\right)^k$, where $\rho(t)\approx\rho_0+\left(\frac{d\rho}{dt}\right)_0(t-t_0)$. Here ℓ and ρ_0 are constant parameters with the physical essences of the initial distance and dark matter density. As for the index of dark matter density's profile k it was specified as -1, 0, and +1.

In spite of the numerical minuteness of these effects (for example $\Delta\omega \sim 10^{-25}\,Hz$) just at the upper magnitude of dark matter density $\rho_0 \sim 10^{-16}\,g/cm^3$ their possible physical application was aimed to improve the astronomical unit.

In this subchapter the effect of light rays refraction in the dark matter's halo and its profiles is considered and the corresponding Cherenkov radiation is searched. In conclusion the powers of such radiation for optical, ultraviolet and X-ray frequency diapasons were estimated and the possibility of their detection is also discussed.

The basic morphological elements of a galaxy, as it was mentioned above, are the nucleus, bulge, gaseous-dust disk, clouds of neutral hydrogen, star clusters and galactic halo. Here we'll focus our attention on the galactic halo only, because it contains most of the galaxy's mass and consists of mostly dark matter [26, 27].

For simplicity we start from the spherical-symmetric model of the dark matter halo. The 4-dimensional interval of the corresponding gravitational field can be written down in the standard form [28]

$$ds^2 = e^{\nu(r)} \cdot c^2 dt^2 - e^{\lambda(r)} \cdot dr^2 - r^2\left(d\theta^2 + \sin^2\theta \cdot d\varphi^2\right), \tag{3.1}$$

where

$$e^{\lambda(r)} = \left(1 - \frac{8\pi G}{c^2 r}\int_0^r \rho(r) r^2 dr\right)^{-1} \tag{3.2}$$

and

$$e^{\nu(r)} = \exp\int_r^\infty \left[\frac{8\pi G}{c^4}\left(c^2\rho(r) + p(r)\right)r \cdot e^{\lambda(r)} - \frac{d\lambda}{dr}\right]dr. \tag{3.3}$$

In expressions (3.2) and (3.3) $\rho(r)$ and $p(r)$ are the mass-energy density and pressure of dark matter, accordingly. The real expressions for mass-energy density and pressure are possible to get by searching the rotational curves of galaxies. Meanwhile, the plausible space distribution for these expressions is possible to get using the N-body simulation method. In fact, the profiles for dark matter energy [29], temperature profiles [30] and some other characteristics of dark matter have resulted from this method.

In the following, as the current variable $\rho(r)$ in (3.2) – (3.3) we'll choose two profiles: Navarro-Frenk-White [2] and Burkert [3], only. Concerning the pressure, we put that it equals zero, as before (see [31], also).

So, for the Navarro-Frenk-White profile $\rho(r) = \rho_0 \Big/ \frac{r}{r_0}\left(1 + \frac{r}{r_0}\right)^2$ the gravitational field's refractive index is (see (1.14))

$$n(r) \approx 1 + \frac{4\pi G}{c^2}\rho_0 r_0^2\left(1 + \frac{r}{r_0}\right). \tag{3.4}$$

Analogously, for the case of the Burkert profile $\rho(r) = \rho_0 \Big/ \left(1 + \frac{r}{r_0}\right)\left(1 + \frac{r^2}{r_0^2}\right)$ the expression of the refractive index has the form (see (1.23))

$$n(r) \approx 1 + \frac{4\pi G}{c^2} \rho_0 r_0^{\,2} \left(\frac{r}{r_0} \right). \tag{3.5}$$

Next, the power of Cherenkov radiation at a frequency ω is described by the expression [17, 18]

$$\frac{d\mathrm{E}}{dt} = W = -\frac{q^2 v}{c^2} \int_{\omega} \left(1 - \frac{c^2}{n(\omega)^2 v^2} \right) \omega d\omega , \tag{3.6}$$

where $n(\omega)$ is the dispersion refractive index, q is the total charge of particles that moves with mean velocity v. As in our case, the refractive index depends on current coordinate r, but not on a frequency (the dispersion absence) as expression (3.6) simplifies

$$\frac{d\mathrm{E}}{dt} = W = -\frac{q^2 v}{2c^2} \left(1 - \frac{c^2}{n(r)^2 v^2} \right) \omega^2 . \tag{3.7}$$

That is why the last expression be applied to (3.4) and (3.5) and take on the forms

$$\frac{d\mathrm{E}}{dt} = W_{NFW} \approx -\frac{q^2 v}{2c^2} \left(1 - \frac{c^2}{v^2} + \frac{8\pi G}{v^2} \rho_0 r_0^{\,2} + \frac{8\pi G}{v^2} \rho_0 r_0^{\,2} \left(\frac{r}{r_0} \right) \right) \omega^2 , \tag{3.8}$$

$$\frac{d\mathrm{E}}{dt} = W_B \approx -\frac{q^2 v}{2c^2} \left(1 - \frac{c^2}{v^2} + \frac{8\pi G}{v^2} \rho_0 r_0^{\,2} \left(\frac{r}{r_0} \right) \right) \omega^2 . \tag{3.9}$$

3.2. Numerical Estimation of the Cherenkov Radiation Power in the Halo of Dark Matter

For the following analyses of the gravitational analog of Cherenkov radiation we have to attract the well-known mechanisms of charged particle creation in a galaxy and to estimate the magnitude of probable charge that comes up to the Earth.

First of all consider the possibility of Sun radiation and radiation from Wolf-Rayet stars usage. It is known that sun wind – the flux of ionized particles (protons, electrons and α -particles) - flows from the solar corona. The slow sun winds have the velocity $v \sim 4\cdot 10^5\ m/s$, while the velocity of fast sun wind is $v \sim 7\cdot 10^5\ m/s$.

The average Sun emits about $1.3\cdot 10^{36}$ particles per second [32] due to the wind. In spite of the vast number of particles, this type of radiation couldn't be used for the estimation of Cherenkov radiation due to the fact that their velocities are radically smaller with respect to the speed of light and due to the minuteness of their flux.

However, the energy diapason of the solar cosmic rays at Sun flare is about $10^5 \div 10^{11} eV$ that corresponds to their maximal velocities order of $10^6\ m/s$. Moreover, in this case the essentially stronger and cosmic ray flux - its magnitude may be $\dot{\eta} = 10^6 cm^{-2} \sec^{-1}$ of order. But the dark matter does not influence the propogation of cosmic rays because the region of their movement is limited by the particular size of the Solar system.

More active in the aspect of the star's wind emission are the Wolf-Rayet stars. Their star's wind emits about three Sun masses per one million years. Therefore the rate of substance loss for Wolf-Rayet stars is $\dot{M} \sim 10^{20}\ g/s$ that is seven orders larger than the rate of a substance loss for the star's wind. Beside, according to [33] the brightness for such type of stars is $10^5 - 10^6$ times larger than the brightness of the Sun. That is why the velocities of the ionizing particles from these stars are essentially larger than from the quiet Sun and enrich magnitudes of about $10^7\ m/s$.

Moreover, particles in the galactic space undergo the additional acceleration by shock waves to acquire the high and ultrahigh energies $10^{15} \div 10^{19} eV$ of order [34]. That is why their final velocities $v \sim 10^7\ m/s$ also compare with the speed of light. And though the number of such particles is very small - one particle per meter squared during one year on the Earth - they may create the Cherenkov radiation at their traverse through the dark matter in principle.

Examination of (3.8) and (3.9) leads to the significant conclusion that from the physical viewpoint the most interesting is the first expression because it includes the largest term $\frac{8\pi G}{v^2}\rho_0 r_0{}^2$ depending on the central magnitude of dark matter density ρ_0. That is why below we'll focus our search on this term in (3.4) and estimate the magnitude of it, only.

To do this we must set the density ρ_0 and take into account the following natural restriction $\frac{8\pi G}{v^2}\rho_0 r_0{}^2 < 1$. Note that in article [25], mentioned above, the upper magnitude of cold dark matter's density is approximately $\rho \sim 10^{-16}\, g/cm^3$. But such magnitude seems too large because with the restriction given above it must be $\rho_0 < 10^{-18}\, g/cm^3$. Remember that the mean magnitude of dark matter's density in the Universe is $\rho \sim 0.3\Omega \sim 10^{-30}\, g/cm^3$. Hence, let the following interval for the dark matter's density takes place

$$10^{-30}\, g/cm^3 \le \rho_{DM} < 10^{-18}\, g/cm^3 . \tag{3.10}$$

As $\frac{G}{c^2} \sim 10^{-28}\, cm/g$, the sizes of the typical galaxy are $r_0 \sim 50\cdot 10^3\, pc \sim 5\cdot 10^{22} cm$, then setting $\rho_0 \sim 10^{-18}\, g/cm^3$ we get the following interval for the above-mentioned term

$$10^{-12} \le 8\pi \frac{G}{c^2}\rho_0 r_0{}^2 < 10^0 . \tag{3.11}$$

Further from (3.4) it follows that the velocity of a charged particle has the expression

$$v \approx c\left(1 - \frac{4\pi G}{c^2}\rho_0 r_0{}^2\right). \tag{3.12}$$

Hence the interval of its numerical magnitudes is

$$c\left(1-10^{-12}\right)\leq v\leq c\,. \tag{3.13}$$

According to the argument mentioned above, we figure that in the halo of dark matter the charged particles can also be accelerated up to the velocities $v\sim 10^{7}\ m/s$. Thus they will correspond to velocities of ionized particles from the Wolf-Rayet stars and satisfy the relation $\frac{v}{c}\sim 10^{-1}$. From these statements we get the magnitude of the central part of the dark matter's density

$$\rho_0=\frac{1-10^{-1}}{\frac{8\pi G}{c^2}r_0^{\,2}}\sim 10^{-19}\ g/cm^3\,. \tag{3.14}$$

This estimation at three orders is smaller than the corresponding value argued in [25]. But the last value, as it seems, is overvalued because the charged particle's velocity will tent to the speed of light at the density $\rho=\rho_0\sim 10^{-16}\ g/cm^3$. Accordingly, this causes the overvaluing and the magnitude of the dark matter's density central part $\rho_0\sim 10^{-14}\ g/cm^3$ that was argued in [35]. (Here we are not concerned with the ultra light particles existence [18] and the results of novel experiment on their detection [36].)

Now based on these data we can estimate the possible magnitude of the Cherenkov radiation in the halo of dark matter for the different (visible, ultraviolet and X-rays) frequency diapasons.

Putting that the flux of high-energy charged particles is uniform and isotropic it is possible to estimate the charge's quantity that falls down on the Earth during one year. Setting that the Earth's radius is $R_E\sim 10^{9}\,cm$, the area of its hemisphere is $S_E\sim 10^{19}\,cm^2$, the rate of particles changing is $\dot{N}=\dot{\eta}S_E\sim 10^{25}\sec^{-1}$. Next, a charge particle moving with the velocity $v\approx 10^{7}\ m/s$, which is close to the speed of light, will cross a galaxy during $t_0=\frac{r_0}{v}\sim 10^{15}\sec$. Hence, the number of upcoming charge particles will be $N=\dot{N}\cdot t_0\sim 10^{40}$ and the corresponding entirety charge equals $Q=eN\sim 10^{21}C$.

After that, assuming that the area of a Cherenkov telescope mirrors lies in the limits $s_{ChTel} \sim \left(10^7 \div 10^{13}\right) cm^2$ (that is identical to increasing such telescope's sensitivity [37]), it is possible detect a charge magnitude of about $q_{\max} \sim 1 \cdot C$.

As we are interested in the influence of dark matter on the Cherenkov radiation then the sum of the two first terms in (3.8) at velocities close to the speed of light will tend to be zero. Hence, the look for power radiation is

$$-\frac{d\mathrm{E}}{dt} = W_{NFW} \approx 4\pi G \frac{q^2}{c^2 v} \rho_0 r_0 \omega^2 . \qquad (3.15)$$

Putting here the orders of frequencies that relate to the visible ($\omega \sim 10^{15} \sec^{-1}$), the ultraviolet ($\omega \sim 10^{17} \sec^{-1}$) and the X-ray ($\omega \sim 10^{19} \sec^{-1}$) diapasons we get the following set of power estimations $\left(1\frac{C^2}{cm} \sim 10^1 GeV \sim 10^{-3} erg\right)$ for the one coulomb charge –

$$\begin{aligned}
-\frac{d\mathrm{E}}{dt} &= W_{NFW} \sim 10^{-1}\, GeV/\sec, \\
-\frac{d\mathrm{E}}{dt} &= W_{NFW} \sim 10^{3}\, GeV/\sec, \\
-\frac{d\mathrm{E}}{dt} &= W_{NFW} \sim 10^{7}\, GeV/\sec,
\end{aligned} \qquad (3.16)$$

accordingly.

As it seems the largest power of the Cherenkov radiation is for the X-rays diapason. That is why the energy order of $\mathrm{E} \sim 10^{20} GeV$ will emit during the particles' time crossing a galaxy $t_0 = \frac{r_0}{v} \sim 10^{13} \sec$. Hence, the corresponding energy density in a halo will be $\varepsilon \approx \frac{\mathrm{E}}{r_0^{\,3}} \sim 10^{-47}\, GeV/cm^3 \sim 10^{-44}\, erg/cm^3$.

Finally, based on the Stephan-Boltzmann law $\varepsilon = \sigma T^4$, where $\sigma \sim 10^{-15}\, erg/cm^3 \cdot K^4$ [38], it is possible estimate the temperature of

Cherenkov radiation. Its largest order (X-ray diapason) is about $T \sim 10^{-6} K$ that is close to the temperature fluctuations of cosmic background radiation $\Delta T \sim 10^{-5} K$ [39]. Hence, it lies on the threshold of modern astrophysical measurements.

Conclusion

What is the physical essence of this type of radiation?

Firstly, its detection may be the criterion of choosing the preferable dark matter distribution in a galaxy. In fact, there are now a number of other types of dark matter profiles in addition to those cited above. They were researched in many articles, for example in [40 – 44], etc. Although all of them describe the structure of dark matter distribution in a galaxy – independent of its red shift, mass, sizes, and character of the evolution – it is desirable to have the general shape of such distribution. It allows the possibility of understanding the universal properties of dark matter.

Secondly, in [15] it was shown that the refracting index for the Einasto [40] dark matter profile is smaller than one. This fact is possible to interpret, probably, as the Cherenkov absorption in the dark matter. Hence, we get the additional perspectives for the dark matter search because up to the present time it was considered that dark matter affects only gravitationally, and the absorption processes in it searched only for the neutrino [45].

Concerning the above type of radiation it is necessary to mention some other articles.

So, in article [46] it was shown that high-energy particles are traversing the Universe through the cosmic microwave background radiation and can, in principle, emit the Cherenkov radiation. It was shown that the energy threshold for this radiation is extremely high and its intensity would be too low due to the low density of the “relic photons gas” and very weak interaction of two photons.

In article [47] this effect was also searched in the relic radiation field. For this medium the refracting index was calculated in the framework of quantum electrodynamics. Its difference from the vacuum-refracting index ($n=1$) is about $\delta \sim 10^{-42}$ and is drastically smaller than in our case - $\delta_{\min} \sim 10^{-13}$.

So, the Cherenkov radiation at the cosmological scales is preferable for searching the galaxy’s dark matter halo.

The possibility of searching dark matter properties using Cherenkov telescopes was recently discussed in article [48].

And last - the electronic publication of the matters in this article have been noted in the reference [49].

4. On the Dark Matter's Halo Theoretical Description

Introduction

The outstanding peculiarity of modern cosmology consists in that it allows us to carry out the high-precision measurements of the physical parameters of the Universe that were considered impossible not so long ago. Talk, for example, about the measurement of cosmic microwave background anisotropy [50], about the polarization of cosmic microwave backgrounds [51], about the gravitational microlensing [52] among other observable effects of modern cosmology.

This peculiarity of modern cosmology allows the broad use of observable data for the next development of theoretical models; those processes that are paramount for understanding the structure and evolution of the Universe. In doing this we concentrate our attention on a problem that is orientating on the understanding of the dark matter phenomena. Some new results in this sphere have been presented recently in the article [53].

This subchapter is organized as follows. In the first subsection I briefly consider the scalar field theory for describing a galaxy's halo of dark matter. In the second subsection I consider the scalar field of the oscillator type and demonstrate that such potential couldn't be productively used for acquiring adequate galaxies' curve lines and determining the profiles of their dark matter halo. To do this it is necessary to choose another type of scalar potential. This problem was researched in subsection 3. There it was argued that global properties of scalar fields is possible to describe by the central symmetric potential of celestial – mechanical type, i.e. by $U(\varphi) = -\frac{\mu}{\varphi}$. Subsection 4 was devoted to calculating profiles of the dark matter halo and finding rotational curves of a galaxy based on celestial – mechanical type potential. Some astronomical predictions concerning the physical characteristics of dwarf galaxies were argued in the conclusion of this subsection.

4.1. Scalar Field for Describing the Halo of Dark Matter

In discussing dark matter it's necessary to mention the pioneer Zwicky article [54], where the idea about the presence of an unknown type of substance (dark matter) in galaxies was first put forward. This substance ensures the stability of any galaxy. That is why the modern vision about the galaxy structure includes the dark matter halo as its indispensable component [55].

Accordingly, this is the simplest model of a galaxy the following system of number subsystems is considering – massive nuclei (the rotating black hole, usually), bulge, spherical stars shell, gaseous flat disc and halo of dark matter. Note that stars shell and flat disc have common sizes, in order of $10Kpc$, while the typical sizes of dark matter halo are about of $100Kpc$ and larger. (Such structure of a galaxy was used in the second and third subchapters of this chapter.)

The observable data show that the halo of dark matter contains the main part of galaxy's mass – about 90%. The typical total galaxy mass is estimated as $M_{Gal} \sim (10^{12}\text{-}10^{14}) M_{\odot}$, where $M_{\odot}$ - mass of the Sun [56]. Because the goal of the presented subchapter is searching some properties of the dark matter, only, we'll neglect all of galaxy's components except the halo of dark matter.

In accordance with the number of articles, enumerated, for example in [55], it is possible to describe the dark matter by setting the suitable scalar field. For classical scalar field φ the energy-momentum tensor has the form

$$T_{\mu\nu} = \partial_{\mu}\varphi \cdot \partial_{\nu}\varphi - \frac{1}{2} g_{\mu\nu} \left[(\partial_{\alpha}\varphi)^2 - U(\varphi) \right], \tag{4.1}$$

where $U(\varphi)$ - the potential energy of the scalar field, $g_{\mu\nu}$ - the metric tensor of an external gravitational field. Later on we'll consider the flat space-time, i.e. $g_{\mu\nu} = \delta_{\mu\nu}$.

Based on (4.1) it is easy get the expression for the scalar field energy density

$$T_{00} = \rho = \frac{1}{2} \left[(\partial_m \varphi)^2 + U(\varphi) \right], \tag{4.2}$$

the space momentum tensor

$$T_{kl} = \partial_k \varphi \cdot \partial_l \varphi + \frac{1}{2}\left[\left(\partial_m \varphi\right)^2 - U\left(\varphi\right)\right] \cdot \delta_{kl} \tag{4.3}$$

and the expression for isotropic pressure $p = \frac{1}{2} \cdot \left[\left(\partial_m \varphi\right)^2 - U\left(\varphi\right)\right]$.

Now we'll find the field equation by using the law of energy-momentum conservation

$$\nabla_v T^{\mu v} = 0\,. \tag{4.4}$$

Remembering that $g_{\mu v} = \delta_{\mu v}$ and considering the Newtonian limit, from (4.4) we get

$$\Delta\varphi + \frac{\partial U\left(\varphi\right)}{\partial \varphi} = 0\,. \tag{4.5}$$

4.2. Scalar Field of the Oscillator Type

Choosing the effective potential energy as $U\left(\varphi\right) = \frac{1}{2} m^2 \varphi^2$ that corresponds to the standard scalar potential of the oscillator type, we get the well-known equation

$$\Delta\varphi + m^2 \varphi = 0\,. \tag{4.6}$$

Passing to the new variable $\psi = \varphi \cdot r$ and using the spherical coordinates, (4.6) rewrites as

$$\frac{d^2 \psi}{d\tilde{r}^2} + \psi = 0\,, \tag{4.7}$$

where $\tilde{r} = mr$ is the dimensionless variable. It has the standard solution of oscillator type

$$\psi = \psi_0 \exp\left(i\tilde{r}\right) = \psi_0 \exp\left(imr\right). \tag{4.8}$$

Now the question arises – of what magnitude the variable $\tilde{r} = mr$ may be? For its estimating remember that minimal mass m of the scalar field particles, describes the cold dark matter, according [57] equals $10^{-23} eV$. Then in usual units ($1eV \approx 1.8 \cdot 10^{-39} g$; $\hbar \approx 1,0 \cdot 10^{-27} erg \cdot \sec$; $c \approx 3,.0 \cdot 10^{10} cm \cdot \sec^{-1}$) the maximal corresponding distance will be

$$r_{\max} \approx \frac{\hbar}{c} m^{-1} \sim 10pc = 0,1Kpc \,. \tag{4.9}$$

But from the set of observable data the most interesting distances for a galaxy's curve lines is about ten and more kilo parsecs. Hence, $\tilde{r}$ is the parameter that satisfies following condition $\tilde{r} = mr = \dfrac{r}{r_{\max}} >> 1$.

Substitution (4.8) and its derivative into (4.2) gives the explicit and exact form of dark matter density profile

$$\rho(r) = \frac{\psi_0{}^2}{2\mathrm{r}^2}\left[\left(\frac{1}{r^2} - \frac{1}{r_{\max}{}^2}\right) + \frac{1}{r_{\max}{}^2}\right]\exp(2imr) \,. \tag{4.10}$$

From expression (4.10) anyone easily gets the scalar field mass density

$$\rho(r) = \frac{m^2 \psi_0{}^2}{2} \cdot \frac{\cos^2 mr}{m^2 r^4} = \rho_0 \cdot \frac{\cos^2 mr}{m^2 r^4} \,. \tag{4.11}$$

From all of above we may write down the Poisson equation in spherical coordinates for searching the gravitational potential $\Phi(r)$ produced by the scalar field

$$\frac{1}{r^2} \cdot \frac{d}{dr}\left(r^2 \frac{d\Phi(r)}{dr}\right) = -4\pi G \rho_0 \cdot \frac{\cos^2 mr}{m^2 r^2} \,. \tag{4.12}$$

After its first integrating and omitting the periodical terms we get the following functional dependency

$$\frac{d\Phi(r)}{dr} \sim -\frac{1}{r^3}, \tag{4.13}$$

after second integrating we find the potential dependency on distance

$$\Phi(r) \sim -\frac{1}{r^2}. \tag{4.14}$$

Expression (4.13) allows us to find the rotational curve line by equating it to the specific centrifugal force, i.e. $\frac{v^2}{r} \sim \frac{1}{r^3}$. So, the rotational curve line has the next functional dependency

$$v(r) \sim \frac{1}{r}. \tag{4.15}$$

Now the question arises - how does the curve line (4.15) correlate with the real observable data? Basing on the observable results of dark and baryonic matter distributions in 34 bright spiral galaxies [58], we may conclude the following.

First, there are a number of variants that model the density distribution of dark matter in halos using the N-body simulation method [2, 3, 5]. Most of them satisfactorily predict that dark matter begins to play the essential curved role for that regions of galaxies, where $\frac{r}{r_{\max}} \geq 1$, in contradiction to maximal sizes of the halo (4.9) which we considered previously.

Second, the observable rotational curves aren't represented by the lines type of hyperbola. Hence, expression (4.15) is the result of poor theoretically determined dark matter distribution even in galaxies where the presence of dark matter is dominant.

From all said above, it is clear that standard scalar potential couldn't be productively used for getting the adequate galaxies' curve lines and determining the profiles of dark matter halo. To do this it is necessary to choose another type of scalar potential. Note that the Yukawa-type potential for describing dark matter has recently been considered by Loeb and Weiner [59].

4.3. Scalar Field of the Celestial – Mechanical Type

In fact, the potentials type of $U(\varphi)=\frac{1}{2}m^2\varphi^2$ and analogous them (self-acting potentials, Higgs potentials, Yukawa-type potential, etc.) are regularly used to search the local properties of the scalar fields on small space and time intervals (very early and early Universe, for example). But our aim is discovering the global properties of scalar fields that can describe the observable characteristics of dark matter on the cosmological space and time scales.

The global properties of scalar fields, as it seems, is possible to describe by the simplest form of central symmetric potential celestial – mechanical type, i.e. by $U(\varphi)=-\frac{\mu}{\varphi}$. Thus the corresponding unit Lagrangian takes on the form

$$L=\frac{1}{2}\left(\vec{\nabla}\varphi\right)^2+\frac{\mu}{\varphi}, \qquad (4.16)$$

where μ is any constant value. Its possible interpretation sees the conclusion to this subsection. For deducing (4.16) we based on the Lagrangian a probe particle that moves in the central symmetric field and used the standard formal replacing $t\rightarrow x^k$ and $r=\sqrt{\left(x^k\right)^2}\rightarrow\varphi$, that usually applies in the field theory [60].

Earlier it was pointed out that the simplest halo of dark matter has the spherical – symmetric shape, i.e. it depends on radius r only. That is why we consider the one-dimensional operator nabla $\vec{\nabla}\rightarrow\frac{d}{dr}(\ )\cdot\vec{e}_r$. So, in place of (4.6) we get the following equation in the dimensionless values r and φ (as in (4.7))

$$\frac{d^2\varphi}{dr^2}+\frac{\mu}{\varphi^2}=0, \qquad (4.17)$$

that describes the linear "movement" of a unit mass particle in the central field. The angular momentum M for such type of "movement", as it is well known, equals to zero, i.e. we may set $M = 0$. Hence, for finding $\varphi(r)$ and its first derivative $\frac{d\varphi(r)}{dr}$ it prefers use of the Lagrangian (4.16) rather than the equation of "motion" (4.17). According to classical textbooks [61] we have

$$r = \int \frac{d\varphi}{\sqrt{2\left(E + \frac{\mu}{\varphi}\right)}}. \tag{4.18}$$

Now consider the field analogs of some types of movements that are interesting from a dynamical viewpoint and have the width applications in celestial mechanics.

i) Assume that "total energy" is larger than "potential energy", i.e. that $E >> \frac{\mu}{\varphi}$. From the physical viewpoint this condition means choosing the space region of the dark matter halo that is close to the center of a galaxy. Then integral (4.18) takes on the approximate expression (integrating constant was included into r)

$$r = \frac{1}{\sqrt{2E}}\left(\varphi - \frac{\mu}{2E}\ln\varphi\right). \tag{4.19}$$

For its inversing let $r = r_0 + \delta r$ and $\varphi = \varphi_0 + \delta\varphi$, where r_0 and φ_0 are the main terms, while δr and $\delta\varphi$ are the small additives order of $\frac{\mu}{E\varphi_0} << 1$ to them. That is why $\varphi_0 = \sqrt{2E \cdot r_0}$ and $\delta\varphi = \sqrt{2E} \cdot \delta r + \frac{\mu}{2E}\ln\varphi_0$. These expressions allow us to write down the final result in the following form and with the above-mentioned accuracy

$$\varphi(r) = \sqrt{2E} \cdot r + \frac{\mu}{2E}\ln\left(\sqrt{2E} \cdot r\right). \tag{4.20}$$

ii) Let the opposite correlation $E << \frac{\mu}{\varphi}$ take place. This condition, contrary to the previous one, means choosing the space region far from the center of the galaxy. Then from (4.18) we immediately get (integrating constant was included into r , also)

$$r = \frac{1}{\sqrt{\mu}} \left(\frac{2}{3} \cdot \varphi^{3/2} + \frac{1}{2} \frac{E}{\mu} \cdot \varphi^{5/2} \right). \tag{4.21}$$

In full analogy with the first variant we set $r = r_0 + \delta r$ and $\varphi = \varphi_0 + \delta\varphi$, where r_0 and φ_0 are the main terms, whereas δr and $\delta\varphi$ are the small additives in order of $\frac{\mu}{E\varphi_0} >> 1$ to them, now. Such representation leads to the next approximate functional dependency

$$\varphi(r) = (\frac{3}{2}\sqrt{\mu})^{2/3} \cdot r^{2/3} + \frac{1}{2} \cdot (\frac{3}{2})^2 \cdot \frac{E}{\sqrt[3]{\mu}} \cdot r^{4/3} . \tag{4.22}$$

So, expressions (4.20) and (4.22) explicitly describe the scalar field potential in the whole region – close to the center and far from the center – of dark matter halo distribution.

4.4. Profiles of Dark Matter Halo and Rotational Curves of a Galaxy

The profile of dark matter is the important characteristic of a galaxy structure. It allows us to calculate the corresponding gravitational potential, find the galaxy's rotational curve and make some other cosmological conclusions. In most articles the profiles of the dark matter halo and the rotational curves of a galaxy were searched mainly by the N-body simulation method. But we'll consider them from the theoretical field viewpoint.

Now, based on (4.20) and (4.22), it is possible to find the profiles of the dark matter halo and rotational curve lines for any galaxy.

i) First of all it is easy to calculate the first derivatives from potential (4.20) -

$$\frac{d\varphi(r)}{dr} = \sqrt{2E} + \frac{\mu}{\sqrt{2E}} \cdot \frac{1}{r}. \tag{4.23}$$

Substituting (4.20) and (4.23) into (4.2) we get the following mass density profile of dark matter

$$\rho(r) = \frac{1}{2}\left[\left(1 + \frac{1}{\sqrt{2E}}\right)\frac{\mu}{r} - \left(\frac{\mu}{2E}\right)^2 \frac{1}{r} \ln\sqrt{2E} \cdot r\right]. \tag{4.24}$$

The integrating constant has been included into the left side of (4.24). Examining it we see that mass density of dark matter decreases in space and becomes equal to zero at the distance

$$\sqrt{2E} \cdot r_0 = \exp\left((2E)^2\left(1 + \frac{1}{\sqrt{2E}}\right)/\mu\right). \tag{4.25}$$

At distances $r > r_0$ the mass density of dark matter becomes growth again. This circumstance is very important, because in the neighboring region the rotational curve line must possess a small "gap" making it possible to observe the principle.

Results of articles [62] gives that authors' mean radial density profile at small distances decrease as $\rho(r) \sim \frac{1}{r}$, which is in good correlation with our estimation (4.24).

Now remembering that ratios $\frac{\mu}{E\varphi_0} << 1$ and $\frac{\mu}{E^{3/2}} \frac{1}{r} << 1$ take place, it possibly omits the second term in (4.24) as the value of a larger order of minuteness. So, our next step is finding the gravitational potential of the field produced by the main term in mass density (4.24).

For this case the Poisson equation takes on the form

$$\frac{1}{r^2}\frac{d}{dr}\left(r^2 \frac{d\Phi(r)}{dr}\right) = -2\pi G\left(1 + \frac{1}{\sqrt{2E}}\right)\frac{\mu}{r} = -2\pi G\frac{\tilde{\mu}}{r}. \tag{4.26}$$

Integration (4.26) gives

$$\frac{d\Phi(r)}{dr} = -2\pi\, G\tilde{\mu} = C = const \tag{4.27}$$

and

$$\Phi(r) = C \cdot r\,, \tag{4.28}$$

hence.

From (4.27) it is easy to get the rotational curve line by equating it to the centrifugal force that acts at a "probe" star. Thus, the following relation

$$\frac{v^2}{r} = C \tag{4.29}$$

and the corresponding simplest rotational curve line

$$v(r) \sim r^{1/2}\,, \tag{4.30}$$

that analogously (4.15) is the direct consequence of (4.29), take places. From the geometrical viewpoint this curve in coordinates $\{v;r\}$ represents the line that slowly grows with increasing distance. That is why the shape of line (4.30) is in good correlation with the real rotational curve lines, because starting from any distance all of them possess a weak trend for growth with the increasing distance.. The results of dark matter's profiles observed at short distances (~ $30Kpc$), are proof of our conclusion, which is presented in article [58].

ii) In full analogy with the previous variant we calculate the first derivative of (4.22)

$$\frac{d\varphi(r)}{dr} = \frac{2}{3}\left(\frac{3}{2}\sqrt{\mu}\right)^{2/3} \cdot r^{-1/3} + \frac{2}{3}\left(\frac{3}{2}\right)^2 \frac{E}{\sqrt[3]{\mu}} \cdot r^{1/3}\,. \tag{4. 31}$$

Substituting (22) and (31) into (2) and remembering that $\frac{E\varphi_0}{\mu} << 1$, analogous to our foregoing result, the approximate expression takes place

$$\rho(r) = \rho_0 + \left[\frac{2}{9}\left(\frac{3}{2}\sqrt{\mu}\right)^{4/3} + \frac{2\tilde{\mu}}{\left(\frac{3}{2}\right)^2} \frac{1}{\frac{E}{\sqrt[3]{\mu}}} \right] \cdot r^{-2/3} = \rho_0 + \sigma_0 \cdot r^{-2/3}, \quad (4.32)$$

where ρ_0 is the corresponding small (order of $\frac{E\varphi_0}{\mu} << 1$) constant value.

Examining the article [62], cited above, we see that the author's mean radial density profile at large distances, contrary to our result (4.32), decreases more rapidly, namely as $\rho(r) \sim \frac{1}{r^{5/2}}$. The same conclusion follows from article [58], where the mass density profile changes according to the dependency $\rho(r) \sim \frac{1}{r^{31/9}}$.

But at the same time the dark matter profile (4.32) is in good correlation with the results of article [63], where the shape of profile density for dark matter dominated the dwarf and low-surface brightness late-type galaxies described as $\rho(r) \sim \frac{1}{r^{\gamma}}$, $\gamma \approx 0.2 - 0.4$. It is important to underline that instead of other articles the power index in profile density for such type of cosmic objects is small.

Now, omitting in (4.32) the constant term we get equality for finding the gravitational potential. In fact, the Poisson equation takes on the form

$$\frac{1}{r^2}\frac{d}{dr}\left(r^2 \frac{d\Phi(r)}{dr}\right) = -4\pi G \frac{\sigma_0}{r^{\frac{2}{3}}}. \quad (4.33)$$

Then

$$\frac{d\Phi(r)}{dr} = -4\pi G\sigma_0 \cdot r^{\frac{1}{3}}. \quad (4.34)$$

From (4.34) we get the rotational curve line by its equating to the centrifugal force. Thus we have the needed relation

$$\frac{v^2}{r} = 4\pi G\sigma_0 \cdot r^{\frac{1}{3}} \tag{4.35}$$

and the corresponding curve line

$$v(r) \sim r^{\frac{2}{3}} . \tag{4.36}$$

From the geometrical viewpoint in coordinates $\{v;r\}$ the curve line (4.36) represents the line that grows with increasing distance more rapidly than curves (4.30). That is why its shape is different from the real rotational curve lines. But it is necessary to remember that the theoretical curve line relates to large distances ($\frac{\mu}{\varphi} >> E$), while all now observable curve lines were plotted for the short distances (about $(30 \div 40)\,Kpc$). And now we are quoting once again the above-mentioned article by Kravtsov and Klypin [64], where the rotational curve reproduces by dependency $v(r) \sim r^q$ with $q \approx 0.9 \div 0.8$. This power index coincides practically with our result (4.36).

Note that behavior of curve lines (4.30) and (4.36), if they apply to galaxies movement, is also in good correlation with the results of searching hidden mass in the Local Group [65].

Conclusion

From the body of the subchapter we may conclude the following. Our choice of scalar field, the celestial – mechanical type, allowed us to productively describe the profiles of dark matter halos at small and large distances. They are in rather good correlation with the results of Kravtsov and Klypin on the modeling of profile density and rotational curve line shapes for the dark matter dominated dwarf and the low-surface brightness of late-type galaxies.

The potential of the oscillator-type, as it is well-known, describes a one-dimensional particle movement, while the celestial-mechanical potential describes a two-dimensional movement in any plane. That is why for the partial case of circular movement along the trajectory $a = const$ it possibly decomposes on two independent orthogonal oscillator-type movements. This statement allows us to find the next relation between mass of particle and parameter μ in expression (4.17) - $m^2 = \frac{\mu}{a^3}$, or in the usual units –

$$m = m_{DM} = \left(\frac{M}{a^3}\right)^{1/2} \cdot \frac{(G)^{1/2} \cdot \hbar}{c^2}. \tag{4.37}$$

For the physical interpretation μ we assume that it is the ordinary celestial-mechanical parameter. Hence $\mu = GM$, where G is the gravitational constant, M is the central gravitating mass.

Based on (4.37) it is possible to calculate masses of particles that describe dark matter around dwarf galaxies, for example. Note that for the spheroid dwarf $M \approx \frac{4}{3}\pi\tilde{\rho} \cdot a^3$, then $\left(\frac{M}{a^3}\right)^{1/2} \approx 2(\tilde{\rho})^{1/2}$. Here $\tilde{\rho}$ may be interpreted as any mean mass density of a dwarf galaxy. As $\frac{(G)^{1/2} \cdot \hbar}{c^2} \approx 8{,}2 \cdot 10^{-52} \cdot cm^{3/2} \cdot g^{1/2}$, then using the modern typical physical characteristics of spheroid dwarf galaxies $\left(M \sim \left(10^7 \div 10^8\right)M_{\odot},\ a \sim 0{,}3Kpc\right)$ [66], we find that $(\tilde{\rho})^{1/2} \sim \left(10^{-10} \div 10^{-11}\right) \cdot g^{1/2} \cdot cm^{-3/2}$. Hence, a particle mass is of order $m_{DM} \sim \left(10^{-22} \div 10^{-23}\right)eV$. This estimation is in very good correlation with the minimal mass of dark matter's particle $m \approx 10^{-23}eV$ [57] that was given before.

Some results of this subchapter have been published in article [67].

Summary

Some new astronomical consequences of the dark matter concept have been considered in detail in this chapter.

In fact, it was shown that the potential of the celestial-mechanical type $U(\varphi) = -\frac{\mu}{\varphi}$ allows us to get rather satisfactory dark matter profiles and rotational curve lines for dwarf galaxies. We pointed out that these results are in good agreement with some previous results, based on the N-body simulation method.

The light rays' propagation is searched in the spherical-symmetric gravitational field of Galaxy described both by dark matter as well as baryonic matter. It is shown that positions of images of the lensing object depend on profiles of dark matter and baryonic matter. Besides, the positions of images from the gravitational lens are drastically smaller than the distance between the gravitational lens and the observer.

The power of the Cherenkov radiation in a galaxy's halo of dark matter, described by profiles of Navarro-Frenk-White and Burkert was searched. It was shown that maximal radiation temperature has the magnitude of about $T \sim 10^{-6} K$.

Moreover, it was pointed out that our observable predictions lay in the threshold of modern astronomical measurements.

Acknowledgments

The author is grateful to the National Space Agency Republic of Kazakhstan for supporting this search in the framework of state program 002 – Space Researches of the Republic of Kazakhstan.

Also many thanks to the director of Virtual Institute of Astroparticle physics (VIA) professor M. Yu. Khlopov for his interest in the gravitational Cherenkov effect and for attracting my attention to some of his prior articles [68 - 72] on the subject matter considered.

References

[1] Chernin A.D. (2001), Cosmic vacuum, *Uspekhi Fizicheskikh Nauk*, 171, 1153–1175.

[2] Navarro J.F., Frenk, C.S., & White, S.D.M. (1995). The structure of cold dark matter halos, ArXiv: astro-ph/9508025; Herritt, D.J., Navarro, E., Ludlow, A., et al. (2005). *Universal density profile for dark and luminous matter.* ArXiv: astro - ph/0502515v1.

[3] Burkert, A. (1995). *The structure of dark matter in dwarf galaxies.*ArXiv: astro-ph / 9504041.

[4] Avila-Reese, V., Firmani, C., Klypin, A. & Kravtsov, A.V. (1999).*The density profiles of dark matter haloes: diversity and dependence on medium.* ArXiv: astro-ph/9906260.

[5] Catena, R., & Ullio, P. (2009). A novel determination of the local dark matter density. ArXiv: astro - ph 09070018v2.

[6] Evans, N.W., & An, J.H. (2005). *Distributions function of dark matter.* ArXiv: astro - ph/0511687v2.

[7] Fock, V.A. (1961). *The theory of space, time, and gravitation.* Moscow: Fizmatgiz, (in Russian).

[8] Zel'dovich, Ya. B. & Novikov, I.D. (1971). *Theory of gravitation and evolution of stars.* Moscow: Nauka, (in Russian).

[9] Drude, P. (1935). *The theory of optics.* Leningrad – Moscow: Gostekhizdat, (Russian translation).

[10] Sparke L.S. (2000). *Galaxies in the Universe: An introduction.* Cambridge University Press, 2000.

[11] Bliokh, P.V., Minakov, A.A. (1989) *Gravitational lenses.* Kiev: Naukova Dumka, (in Russian).

[12] Zakharov, A. F., Sazhin, M.V. (1998) Gravitational microlensing, *Uspekhi phizicheskih nauk,* 41, 945 - 982.

[13] King, I.R. (1962). The structure of star cluster. I. An empirical density law. *Astron. J.,* 67, 471.

[14] Idliss, G.M. (1961), *Structure and dynamics of the stellar systems.* Alma-Ata: NAS Republic of Kazakhstan, (in Russian).

[15] Chechin, L.M., Avkhunbaeva G.M., Aimuratov E.K., Umiralieva A.Zh, (2012), Deviation of light rays in galaxy dark-matter halos. *Russian Physical Journal,* №6, 690 - 698.

[16] Kardashev, N.S. (1983). The phenomenological model of the Galaxy nucleus. In book "*Results of science and technology". Ser. Astronomy.* Moscow, v.24, (in Russian).

[17] Bolotovskij, B.M. (1957). Theoriya effekta Vaviliova - Cherenkova. *Uspekhi Fizicheskikh Nauk*, 62 № 7, 201 – 246.

[18] Ginzburg, V.L. (1975). *Teoreticheskaya fizika i astrofizika*, Moscow: Nauka, (in Russian).

[19] Ivanitzkaya, O.S. (1979). *Lorentzev bazis i gravitatzionnuye effektu v einshtejnovoj teorii tyagoteniya*, Minsk: Nauka i Tekhnika, (in Russian).

[20] Gupta, A., Mohanty, S. & Sama, M.K. (1999). Cherenkov radiation emitted by charged particles in a gravitational field, *Class. Quant. Grav.,* 16, 291 – 298.

[21] Paczyński, B. (1996). Gravitational microlensing in the Local Group, Astro-ph/9604011v1; Zakharov, A.F., & Sazhin, M.V. (1996). Microlensing by noncompact objects, *Pis'ma v ZhETF,* 63, 937 – 943.

[22] Sizun, P., Cassé, M., & Schanne, S. (2006). Continuum gamma-ray emission from light dark matter positrons and electrons, Astro-ph/0607374v2; Premadi, P., Martel, H. R. Matzner et al. (2001). Light

propagation in inhomogeneous Universes II. Cosmological parameter survey, Astro-ph/0101359v1; Moiseev, A., & Proumo, S. (2009). Sources and detection of dark matter and dark energy in the Universe. Proc. 8-th UCLA Symp. 1166, 151; de Boer, W. (2009). Indirect dark matter searches in the light of ATIC, FERMI, EGRET and PAMELA. SUSY09, Northeastern Univ., Boston, June 5.

[23] Gilmore, G., Wilkinson, M.I., Wyse, R.F.G., et al. (2007). *The observed properties of dark matter on small spatial scales*. Astro-ph/0703308v1.

[24] Salucci, P., Nesti, F., Gentile, G., et al. (2010). *The dark matter density at the Sun's location.* Astro-ph.GA/1003.3101v2.

[25] Hideyoshi Arakida, (2008). *Influence of dark matter on light propagation in solar system.* Astro-ph/0810.2827v3.

[26] Helfer, H.L. (2004). The dark matter of galactic halos. *Bull. Amer. Astron. Soc.,* 36, 1393; Clowe, D., Bradaĉ, M., Gonzalez, A. et al. (2006). A direct empirical proof of the existence of dark matter. Astro-ph/0608407v1.

[27] Strigari, L., Koushiappas, S., Bullok, J., et al. (2007). *The most dark-matter-dominated galaxies: predicted gamma-ray signals from the faintest Milky Way dwarfs*. Astro-ph/0709.1519v2.

[28] Zel'dovich, Ya.B., Novokov, I.D. (1975). *Stroyeniye i evolutziya Vselennoij*, Moscow: Nauka, (in Russian).

[29] Rasia, E., Tormen, G., Moscardini, L. (2004). A dynamical model for the distribution of dark matter and gas in galaxy clusters. *MNRAS,* 351, 237 – 252.

[30] Markevitch, M., Forman, W.R., Sarazin, C.L., et al. (1997). The temperature structure of 30 nearby clusters observed with ASCA. *Similarity of temperature profiles.* Astro-ph/9711289v2.

[31] Boehmer, C.G. & Harko, T. (2007). *On Einstein clusters as galactic dark matter halos.* Gr-qc/0705.1756v2

[32] Kallenrod, M.-B. (2004). *Space physics. An introduction to plasmas and particles in the heliosphere and magnetosphere.* Springer; Suess, S. (1999). The solar probe. NASA/ Marshall space flight center, June 3; Carroll, B.W. (1995). *An introduction to modern astrophysics,* 2 ed., Benjamin Cummings.

[33] Barlow, M.J., Smith, L.J., & Willis, A.J. (1981). Wolf-Rayet stars and binarity, *MNRAS*, 196, 101- 110.

[34] Aloisio, R., Beresinsky, V., & Gazizov, A. (2009), Ultra high-energy cosmic rays: *The disappointing model.* Astro-ph.HE/0907.5194v2.

[35] Vasiliev, E., Zelnikov, M. (2008). *Dark matter dynamics in the galactic center.* Astro-ph/0803.0002v2.

[36] Opera Collaboration, (2011). *Measurement of the neutrino velocity with the OPERA detector in the CNGS beam.* Hep-ex/1109.4897.

[37] Actis, M. Agnetta, G., Aharonian, F. et all. (2010). Design concepts for the Cherenkov telescope array. ArXiv:1008.3703v3 (astro-ph.IM).

[38] Gurevitch, L.E., Chernin, A.D. (1978). *Vvedeniye v kosmogoniyu*. Moscow: Nauka, (in Russian).

[39] Smoot, G. F., Scott, D., (1996). *The cosmic background radiation. Astro-*ph/9603157.

[40] Einasto, J. (2000). *Dark matter and large-scale structure.* Astro-ph/0012161v1.

[41] Klypin, A.A. Kravtsov, A.V., Bullok, J., et al. (2000). *Resolving the structure of cold dark matter halos.* II. Astro-ph/0006343v2.

[42] Treu, T., & Koopmans, V.E. (2004). *Massive dark-matter halos and evolution of early-type galaxies to z=1.* Astro-ph/0401373v2.

[43] Weber, M., de Boer, W. (2009). *Determination of the local dark matter density in our Galaxy".* Astro-ph/0910.4272v2.

[44] Kazantzidis, S., Mayer, L., Mastropietro, C., et al. (2003). Density profiles of cold dark matter substructure: *Implications for the missing satellites problem. Astro*-ph/0312194v4.

[45] Esteban, R. (1993). Ultrahigh energy neutrino absorption by neutrino dark matter. *Phys. Rev., D* 47, 5247 – 5251.

[46] Marklund, M., Brodin, G., Stenflo, L., et al. (2005). Cherenkov radiation in a photon gas. *New J. Phys.* 7, No. 70.

[47] Chefranov, S.G. (2010). The Vavilov-Cherenkov radiation by relict photon gas.Astro-ph/1009.0594v1.

[48] Sánchez-Conde, M.A., Cannoni, M., Zandanel, F., et all. (2011). Dark matter searches with Cherenkov telescopes: nearby dwarf galaxies or local galaxy clusters?, *JCAP*, № 12, 011.

[49] Chechin, L.M. (2013). *The Cherenkov radiation in the galaxy halo of dark matter*. ArXiv1302.3683C.

[50] Komatsu, E., Smith, K.M., Dunkley J., et all.(2010). *Seven-year Wilkinson microwave anisotropy probe (WMAP) observations: cosmological interpretation.* ArXiv: 1001.4538v3 (astro-ph.CO).

[51] Larson, D., Dunkley, J., Hinshaw, G., et al, (2010). Seven-year Wilkinson Microwave Anisotropy Probe (WMAP) observations: power spectra and WMAP-derived parameters, Astro-ph.CO/ 1001.4635v2.

[52] Henk, H. & Bhuvnesh, J. (2008). Weak gravitational lensing and its cosmological applications. Astro-ph / 0805.0139v1; Zakharov, A.F. &

Sazhin M. V., (1998). Gravitational Microlensing, *Physics – Uspekhi (Advanced in Physics Science),* 41, 945 – 982.

[53] Doroshkevich, A. G., Lukash, V.N. & Mikheeva, E. V. (2012). A solution of the problems of cusps and rotation curves in dark matter halos in the cosmological standard model, *Physics – Uspekhi (Advanced in Physics Science),* 55, 3 - 18.

[54] Zwicky, F. (1937). On the masses of nebulae and of clusters of nebulae, *The Astrophysical Journal,* 86, 217 – 246.

[55] Clowe, D., Bradač, M., Gonzales A. H. et al, (2006). *A direct empirical proof of the existence of dark matter,* Astro-ph/0608407v1.

[56] de Luca, G., Kaufmann, G., Springel, V., et all, (2003). Substructures in dark matter haloes, Astro-ph / 0306205v2; Moore, B. Ghigna, S., Governato, F., et al, (1999), *Dark matter substructure in galactic halo,* Astro-ph / 9907411v1.

[57] Gorbunov, D.S., & Rubakov, V.A. (2009), *Introduction to theory of early Universe,* Moscow: URSS, (in Russian).

[58] Susan, A. Kassin, R. de Jong, S., et al, (2006), *Dark and baryonic matter in bright spiral galaxies:* II. Radial distributions for 34 galaxies, Astro-ph/0602027v1.

[59] Loeb, A., & Weiner, N., (2010), Cores in dwarf galaxies from dark matter with a Yukava potential, Astro-ph CO/1011.6374v1.

[60] Schmutzer, E., (2072), *Symmetrien und Erhaltungssätze der Physik*, Berlin: Akademie Verlag.

[61] Goldstein, H., Poole, C., & Safko, J. (2002), *Classic Mechanics,* Addison Wesley; L. D. Landau and E. M. Lifshitz, *Mechanics*, 3-ed, (1976), Pergamon Press.

[62] Kazantzidis, S., Mayer, L. & Mastropietro, C. (2003), Density profiles of cold dark matter substructure: implications for the missing-satellites problem, Astro-ph/0312194v4.

[63] Dehnen, W. & McLaughlin, D. E. (2005), Dynamical insight into dark-matter halos, Astro-ph/0506528v1.

[64] Kravtsov, A. & Klypin, A. A. (1997). The cores of dark matter dominates galaxies: theory vs. observations, Astro-ph / 9708176v2.

[65] Karachentsev, I. D. (2001), Hidden mass in the Local Group, *Physics Uspekhi (Advanced in Physics Science),* 44, 817 - 818.

[66] Sholnik, E. L. Liu, M. C., Reid, I. N., et al., (2010*), Searching for young dwarfs with GALEX,* Astro-ph/ 1011.2708v2.

[67] Chechin, L. M., (2012), On the dark matter's halo theoretical description. *Journal of Modern Physics,* 2012, v. 3, 377 – 382.

[68] Zeldovich, Ya. B., Klypin, A. A., Khlopov, M.Yu., et al., (1980). Astrophysical bounds on the mass of heavy stable neutral leptons. *Sov. J. Nucl. Phys.* 31, 664 - 669.
[69] Konoplich R. V., & Khlopov. M. Yu. (1994). Astrophysical constraints on mass of very heavy stable neutrino. *Yadernaya Fizika,* 57, 452 - 458.
[70] Fargion, D., Khlopov, M. Yu., Konoplich R.V., & Mignani R. (1996). Is stable tau neutrino really an allowable cold dark matter candidate? *Phys. Lett.,* 1996, B375, 223 - 228.
[71] Fargion, D., Khlopov, M. Yu., Konoplich R.V., & Mignani R. (1995). Bounds on very heavy relic neutrinos by their annihilation in the galactic halo. *Phys. Rev.,* D52, 1828-1836.
[72] Belotsky, K.M., & Khlopov M. Yu. (2005). On the distribution of collisionless particles in local potential well. *Gravitation & Cosmology,* 11, No. 3 (43), 220 - 222.

In: Research Development in Dark Matter Research ISBN: 978-1-62948-010-7
Editors: Nori Kinjo and Akira Nakajima

Chapter 3

DARK MATTER/ENERGY, QUANTUM MECHANICS AND MODERN CAUSAL ANALYSIS

Michael Joffe
Imperial College London, London, UK

Abstract

Quantum theory was developed at a time when only "ordinary" matter was thought to exist. Its interaction with dark matter or dark energy is similarly absent in the many interpretations of the equations. Moreover, most of these interpretations postulate that reality in the quantum domain, especially causality, operates in a way that seems very strange in relation not only to common sense but also to the other sciences.

The interpretation that was dominant for the first half-century of quantum theory, the Copenhagen interpretation, holds that quantum reality is inherently probabilistic. In parallel to the dual nature of matter, wave-like or particle-like in different settings, a probability distribution that is derived from deterministic causation (Schrödinger's equation) "collapses" to give a single outcome. In addition, a central place is given to measurement in bringing this about. Quantum reality is thus held to be radically different from that in all other domains, including the remainder of science, in which the prevailing view is that nature and its causal relationships exist independently of the observer, i.e., they have their own prior reality. It is however possible to give an account of the subatomic domain that is realist in this sense, and that also reproduces all the predictions of quantum mechanics. This requires taking non-locality seriously, but there is in any case abundant evidence for this being an established feature of the quantum world. The existing theoretical tradition of this type, originating in the work of de Broglie and

Bohm, does not command wide acceptance in physics, but there is scope for it to be developed in a manner that takes account of dark matter and/or energy.

Dark matter was originally identified at the macroscopic level from discrepancies in cosmological measurements and from gravitational lensing, reinforced by additional evidence, e.g., from the Bullet Cluster. At the microscopic level, various types of particle have been proposed as constituting dark matter, including supersymmetric neutralinos or gravitinos, the Kaluza-Klein particle, and axions. The necessary properties of such a particle, aside from gravitational effects, include that it does not interact with electromagnetic radiation such that its macro properties would be affected; there is no proscription of interaction with normal atomic matter. This raises the intriguing possibility that dark matter particles may participate in quantum phenomena. Far less is known about dark energy, and it is possible also that it could interact with normal matter so as to generate the phenomena of quantum physics.

In recent decades, causal analysis has focused on a broad range of scientific topics in disciplines other than physics, notably biology. The nature of causation is now much better understood. Briefly, a causal relationship involves both a mechanistic and a difference-making (e.g., mathematical) aspect, the two aspects being complementary. This perspective also illuminates apparently probabilistic behaviour: rather than being seen as resulting from an intrinsically different type of causation from deterministic behaviour, it reflects multiple/partial causation. In situations as different as coin-tossing and disease risk, the outcome of a conjunction of forces cannot be accurately predicted as it can in the deterministic situation. This is partly a question of level, the individual outcome being determined at the micro level but only visible at the macro level, where it is apparent as a particular probability at the level of a "population" of coin tosses or lung cancer cases. It is also a manifestation of how multiple causes combine to produce a particular outcome, e.g., the way that long-term heavy cigarette smoking causes lung cancer, but only in conjunction with many other causes, most of which are unknown.

If one assumes that the fundamental way the world works is the same in all domains, albeit with some important domain-specific differences, the findings of modern causal analysis can be used to illuminate the role of probabilistic causation in fundamental physics, recasting it as multiple causation. In particular, making the connection between dark matter/energy and quantum theory could help resolve some of the hitherto intractable problems in the latter, while providing new information on the former.

Introduction

Quantum theory is one of the great achievements of twentieth-century physics. The basic mathematics is clear, and generates extraordinarily precise predictions of experimental findings. However, the nature of the physical reality to which it corresponds is controversial. There are numerous interpretations, some of which have considerable followings in the physics profession, but this topic remains unresolved.

The theory was developed on the assumption that the universe was composed only of "ordinary" matter – protons, neutrons, electrons, photons, etc. It is now clear that this forms only a small part of the total mass of the universe, the remainder being dark matter; and in addition it appears that the total mass/energy is largely made up of dark energy. It is in principle possible that a broader perspective could help to resolve the problem of providing a satisfactory physical interpretation of quantum mechanics.

The issues include wave/particle duality; the possibly probabilistic character of the natural world at the deepest level; and the idea that the world has its own existence that is independent of the observer – or alternatively that measurement plays a central role in behaviour at the subatomic level. If the latter is maintained as true, this implies that in a fundamental philosophical sense the way the world works is different in different domains.

Such a broader perspective, by resolving the specific problems of quantum-theory interpretation, could illuminate the properties of dark matter and/or energy while also assisting in clarifying the relationship of quantum theory to relativity theory. At present they are not mutually compatible.

This paper brings together the physics of dark matter/energy and the interpretation of quantum mechanics, together with a discussion of the philosophical issues involved. In particular, insights into the nature of causation, derived from analysis of sciences other than physics, have the potential to illuminate the deeper topics. The hypothesis put forward is that it would be possible to develop a theory that would bring together the nature of dark matter and/or energy, or at least some subset thereof, with a particular interpretation of quantum physics (following from the work of David Bohm), in a manner that fits with the same concept of causation that operates in other domains. It is a hypothesis that this can be done, not the theory itself.

Some work in this direction already exists, of a mathematical nature. Carroll has connected dark matter with the ensemble density characterising the quantum potential, using Weyl geometry [1]. González-Díaz has used the vacuum fields concept of dark energy to present a revised analysis of quantum mechanics; it involves particles that are tachyonic, i.e., they move faster than light [2]. In this paper I focus on the conceptual and physical issues, and exclude consideration of the mathematics.

The paper covers the following topics: key aspects of quantum theory that are relevant to a potential theory of the suggested kind, with special emphasis on non-local phenomena including entanglement and decoherence; a very brief outline of dark matter and dark energy; the philosophy of causation, drawing on modern developments in causal analysis; and finally a section that pulls the threads together.

Quantum Theory

The Probabilistic Nature of Quantum/Subatomic Phenomena

It is well known that the behaviour of matter at the subatomic scale is very different from that seen in macro physics. In particular, probability is central. This is in sharp contrast to classical nineteenth-century (pre-quantum) physics, in which deterministic causation played the dominant role, with equations accurately predicting each individual outcome. Statistics was used only when describing populations of similar elements, as in statistical mechanics applied to the Ideal Gas Law. Whereas probability in classical physics concerned *distributions*, in quantum physics it is *structural* [3].

Descriptively, perhaps the simplest example is radioactive decay, in which the nucleus of an unstable atom loses energy by emitting ionising radiation. The process is well described by an exponential decay function when applied to a large number of events, but the precise moment when emission takes place appears to be completely random.

Connected with inherently probabilistic behaviour is the dual character of quantum-world entities. The basic building blocks of matter, such as electrons and photons, can show either particle-like or wave-like properties, depending on how they are studied. The same is true for larger objects such as neutrons and protons, and even for molecules. Thus, particle-like behaviour is exhibited in such observations as the frequency dependence of the energy of light, and the ability of matter and radiation to be in thermal equilibrium. Wave-like behaviour, on the other hand, is seen in the double-slit experiment in which an interference pattern is seen – a finding which remains valid even when the light intensity is reduced so that only a single photon is emitted.

The wave properties are expressed by Erwin Schrödinger's equation, which describes how the quantum state of a physical system changes over time. It is a deterministic theory that predicts a probability distribution – the probability density of a particular outcome is given by the absolute square of the wave amplitude. In experiments that display the wave property it is impossible to predict the individual outcome, but the *distribution* of outcomes can be predicted to an extremely high level of accuracy by this wave function. In contrast, with an experimental setup that demonstrates any particle-like property, just one particular outcome is observed.

Many interpretations exist that attempt to explain the established mathematical formalism and experimental findings. Here I will not attempt a comprehensive description, but only select the relevant key points.

One interpretation, known as the Copenhagen Interpretation, is that nature is *inherently* probabilistic, and that when a measurement is made of a single outcome, the wave function describing the probabilistic behaviour "collapses" to a single value. This gives a central role for measurement, because it is the act of measurement that causes this alteration in the system that is being observed. In place of the assumption that is commonly made by scientists, that they are studying a reality that actually exists, independent of who happens to be observing it ("realism"), we now have reality depending crucially on measurement. The situation is further complicated because it is not at all clear what measurement actually means – e.g., an apparatus or a human. Werner Heisenberg denied that subjectivism was implied: "The observer has … only the function of registering decisions … it does not matter whether the observer is an apparatus or a human being" [4]. This interpretation has been highly influential, and is widely supposed to reflect the views of the scientists who developed quantum theory in the 1920s, but in fact it appears to have been constructed by Heisenberg somewhat later, and to be at odds with the views of his colleagues, in particular Niels Bohr [5, 6].

Entanglement and Decoherence

An important quantum property is that particles are emitted in pairs that have opposite properties such as their spin or polarisation, a perfect negative correlation that is maintained even after they have become widely separated. Schrödinger introduced the term *entanglement* for this phenomenon, which he said is "... not *one* but rather *the* characteristic trait of quantum mechanics" [7]. Early discussions on this centred around the thought experiment of Albert Einstein, Boris Podolsky and Nathan Rosen (EPR), postulating a local realist theory based on a common cause to explain the appearance of wave-function collapse (see discussion below). There is now abundant experimental evidence for entanglement, showing that it operates very rapidly, and even at distances of the order of 100 km [8, 9].

Since the 1970s, much research has focused on quantum decoherence, the interaction (entanglement) of a quantum system with its environment. According to James Hartle, it is "a more precise, more general, and more observer-independent notion than measurement" [10]. This open system approach contrasts with the ideal in classical physics which is to analyse an isolated system. It is based on Schrödinger dynamics, and retains determinism in the sense that the component of the wave over time can be re-identified, enabling trajectories to be discerned. The decoherence program attempts to explain the emergence of classical behaviour purely from the formalism of basic quantum mechanics, leading to robustness or stability and suppressing interference effects. It is, however, controversial whether decoherence can "solve the problem of definite outcomes in quantum measurement: We are still left with a multitude of (albeit individually well-localized quasiclassical) components of the wave function, and we need to supplement or otherwise to interpret this situation in order to explain why and how single outcomes are perceived" [11]. Furthermore, the boundary of a "system" is unclear, i.e., it can always be enlarged; and the experimental findings on how decoherence operates in different systems imply that generalisations are hazardous [12].

The Possibility of a Realist Theory

There are many other interpretations of the experimental findings and the equations that form the core of quantum theory, discussion of which is beyond the scope of this paper. For example, another influential one is that in place of wave function collapse, the universe repeatedly splits into multiple unobservable alternate histories. Here I will concentrate on a less popular account that focuses on possible hidden variables, which is associated primarily with the names of Louis de Broglie and David Bohm.

The denial of realism in the Copenhagen Interpretation (and others) led to attempts to recast quantum physics so that the act of measurement became less central. In the 1920s, de Broglie put forward an early version of this theory, but was dissuaded from pursuing it further. In the 1950s, Bohm produced his own theory, addressing wave/particle duality by combining a particle theory with a pilot wave, and showed that it was able to produce identical predictions to conventional quantum theory [13]. However, his main intention was "to show ... that alternative interpretations of quantum theory were in fact possible", and he always remained quite tentative about the specific details of his own account [14]. Some consider it the conceptually clearest and simplest theory to explain quantum phenomena [15]. His close colleague Basil Hiley has stated that Bohm's aim was not to set out a deterministic, mechanical viewpoint, but rather to show that it was possible to attribute properties to an underlying reality, in contrast to the conventional approach to quantum mechanics [16].

The EPR thought experiment and the de Broglie-Bohm theory prompted John Stewart Bell to investigate the possibility that a hidden variable theory could reproduce all the predictions of quantum mechanics [17]. He derived an inequality showing that if quantum mechanics is correct, it is necessary to abandon one or both of two assumptions that experiments produce: unique results ("counterfactual definiteness"), and/or the assumption of locality. The EPR hypothesis was therefore untenable. The implication is that a realist theory must involve *non-locality*, which is implicit in many interpretations of quantum theory but is most explicit in Bohm's. It is an idea that Einstein disparaged as "spooky action at a distance", in violation of special relativity. However, non-locality is now a widely accepted property of quantum phenomena, especially in the context of entanglement and decoherence. It can be reconciled with relativity if it does not involve signalling outside the light cone.

Bohm's theory, and others in the same tradition, exist in distinct versions that differ in fundamental respects, e.g., whether or not they are deterministic. No version has commanded widespread acceptance among physicists, although some of the criticisms may have been due to misinterpretations of the theory, e.g., the role of particles [18]. Perhaps the most problematic component is the requirement that non-locality involves the whole universe – the motion of one particle depends on the positions of all of the other particles in the universe [19]. There are additional problems, e.g., that for isolated systems the Bohm theory will typically not give the correct classical limit [20]. It may be that decoherence could help resolve this and other problems with Bohm's theory, e.g., in the selection of quasiclassical particle trajectories [11].

Nevertheless, theories in this tradition have some advantages from a philosophical viewpoint. First, most other interpretations involve a conception of quantum reality that is radically different not only from "common sense" but also from other sciences. Secondly, the realist approach means that particles have definite positions at all times, and the measurement problem is resolved. Collapse is interpreted as phenomenological. Thirdly, it takes place in a single space-time, so does not involve the incessant multiplication of universes inherent in the "many worlds" interpretation – although it could be said that the idea that every particle in the universe affects every other particle is almost as peculiar.

It remains unclear whether a theory is possible that retains these advantages while being a satisfactory account of the physics. If so, it would need to take account of recent work. Gröblacher et al. have shown theoretically and experimentally that any non-local extension of quantum theory has to be highly counter-intuitive, placing a restriction on the possible range of non-local theories [21]. Colbeck and Renner show that no extension of quantum theory can possibly have any additional predictive power, if one assumes the freedom to choose measurement settings, i.e., if the settings are uncorrelated with any pre-existing values in any reference frame [22]. Gisin has shown that Bell's result applies not only to local but also to covariant non-local variables [23].

Dark Matter and Dark Energy

Dark Matter

The existence of dark matter was first postulated in the early 1930s to explain observations that did not fit with the known properties of matter [24]. In 1932, Jan Oort noted that the observed and predicted orbital velocities of stars in the Milky Way galaxy did not coincide, implying that the mass of the galactic plane must be greater than the total mass of visible material. The following year, Fritz Zwicky examined the Coma Cluster, which contains over a thousand galaxies, and similarly concluded that the great majority of its matter was dark, based on measurements of the orbital velocities of galaxies in clusters. Subsequently evidence of several different types has confirmed these initial findings, including gravitational lensing by galaxy clusters such as the Bullet Cluster.

Estimates of the relative abundance of dark matter and "ordinary" matter suggest that the former is five times as plentiful as the latter. Only a small proportion of dark matter is ordinary matter that is in a state whereby it emits little or no electromagnetic radiation, such as neutron stars or black holes; the great majority is "non-baryonic", i.e., unrelated to ordinary matter [25]. It has a characteristic spatial distribution, occurring in roughly spherically symmetric, centrally concentrated halos associated with galaxies, with the visible matter being concentrated in a disc at the centre [26].

There has been much research on the composition of dark matter at the microscopic level. Various types of particle have been proposed [27-30], including supersymmetric neutralinos or gravitinos, the Kaluza-Klein particle, and axions. It is likely that progress will be made on this as a result of current research using the International Space Station. The necessary properties of such a particle, aside from gravitational effects, include that it does not interact with electromagnetic radiation in a manner that would affect its macro properties, although it remains possible that the quantum behaviour of photons is affected if it cancels out at the macro level. There is no proscription of interaction with normal atomic matter.

Dark Energy

Dark energy is a much more recent concept. It originated in the 1990s, following observations indicating that the universe is expanding at an accelerating rate. In particular, redshift measurements showed that Type 1a supernovae are receding much faster from Earth than predicted by Hubble's Law, which states that the velocity is proportional to the distance [31]. It is possible to measure their distance from Earth because they have a known and extremely consistent intrinsic brightness ("standard candles"), so that their observed brightness allows their distance to be measured.

Since then, other evidence has accumulated that point in the same direction. The universe has been shown to be almost flat, based on measurements of anisotropies (direction-dependent properties) in the cosmic microwave background, and this requires that its mass/energy density must be equal to a critical density. Dark energy is not the only possible explanation for these findings, but if it is correct then the implication is that it forms the large majority of mass/energy in the universe, the latest estimate being 71.4 percent, as compared with 24.0 percent dark matter and 4.6 percent ordinary matter [32]. This is corroborated by other types of evidence, e.g., concerning the large-scale structure of the universe [33].

The distinctive characteristic of dark energy is its repulsive nature, referred to as negative pressure, which is responsible for the observed acceleration in the expansion rate of the universe. The two main models are the Lambda-CDM (Λ-CDM, cosmological constant) and quintessence models. In the former, Lambda represents the cosmological constant, sometimes called vacuum energy, a uniform property of the whole universe that can be thought of as the cost of having space. Particle physics predicts the existence of a cosmological constant, but this advantage may be more apparent than real, because quantitatively the predicted value of energy density is greater than the estimated empirical value by 120 orders of magnitude [34]. The Lambda-CDM model is consistent with numerous cosmological observations [35].

Quintessence models propose the existence of a dynamical field, the potential energy of which causes the accelerated expansion of the universe. Unlike the cosmological constant it varies in space and time. No direct evidence for quintessence is currently available.

Modern Causal Analysis

Causation and the Uniqueness of Physics

In recent decades, the philosophy of science has shifted from an almost exclusive focus on physics to a broader perspective on the sciences [36-39]. Biology in particular has received a great deal of attention, and this has resulted in better understanding of major topics including the nature of causation in general, and apparently probabilistic causation in particular. Many fundamental issues are particularly difficult in physics (especially fundamental physics). For example the very existence of causality was questioned by Bertrand Russell in 1912 [40], a view that is still expressed today in the philosophy of physics [41] but not that of the rest of science.

Consideration of these other sciences has shed light not only on their particular circumstances, but also on broader questions that affect science in general. Such insights can be applied now to physics, albeit cautiously and with due attention to its specificity; the assumption that the world works in fundamentally the same way in all domains is clearly not correct in every detail, and this is especially true at the subatomic scale given its behaviour that is so strange from a macro perspective. In this section I will briefly discuss some of these topics: realism; the complementarity of mechanism and difference making (equations etc.), and the role of mathematics; and multiple/partial causation and apparently probabilistic causation.

Realism

The first of these topics, realism, has been a central concern in many of the twentieth century debates about physics, including the interpretation of quantum phenomena. It concerns the status of science's object of study: whether the objects/phenomena have their own existence independent of the observer. Many scientists have a preference for a realist account of science, and it can be argued that scientists investigate their specific piece of nature as if it truly exists, even when they might not explicitly embrace this philosophical position in the abstract [42]. In quantum physics this is controversial for well-known reasons, but it is clear in other sciences.

For example, the study of biology presupposes not only that living organisms and their constituent organs, tissues etc. have their own independent existence, but also that this is true of the causal processes of life. A nerve impulse transmitted across a synapse really does involve a series of causal steps – release of neurotransmitter, its diffusion across the synaptic space, its subsequent stimulation of the post-synaptic neuron, etc. Becoming infected with malaria involves the causal (sic)agent, a protozoan of genus *Plasmodium*, and a particular type of mosquito, and it happens in the same way whether or not an observer is present (unless s/he intervenes).

These statements refer to what actually exists – they are ontic; naturally when one considers how knowledge is obtained – epistemology – the role of the observer does become central. The implication is that if a satisfactory realist quantum theory really is impossible, different rules apply at the deepest level in different sciences, and probably in different parts of physics too.

The Complementarity of Mechanism and Difference Making, and the Role of Mathematics

The philosophy of science literature on causation in recent decades has tended to focus on two aspects. One has come to be called "difference making", a collective term for a number of distinct viewpoints, including counterfactual, probabilistic and manipulationist accounts. This involves diverse interpretations of statistical analysis. The other is mechanism, i.e., how a particular change is brought about, and what it is (physically) that does this [36-38, 43-45].

In recent years, it has increasingly been realised that in reality, causation involves both aspects. Those with a mechanistic approach have begun to incorporate aspects of difference making [44-47] and *vice versa* [48, 49]. The two aspects have been brought together more explicitly in the epistemic view of Federica Russo and Jon Williamson, who showed that scientists apply both mechanistic and difference-making concepts to a single causal claim, not to two conceptually different types of causal relationship [37]. Their solution is epistemic because it states that a rational actor can approach more closely to the aim of an ideal (and unattainable) state of knowledge the more different types of evidence are available to her, including evidence both of mechanistic and difference-making types.

An epistemic account is, however, necessarily incomplete. There must be a source for these various types of evidence, implying that causal relationships in the real world must combine mechanism with difference making. Evidence of the mechanistic type is available because it is possible for science to uncover the structure and mode of operation of the causal mechanism and evidence of the difference-making type is available because the causal mechanism generates differences that can be detected [50]. This can be summed up by saying that acausal relationship is one that has a *mechanism* that by its operation *makes a difference*.

Mechanism and difference making are thus different aspects of a single causal relationship, and they play a complementary role, neatly summarized by John Pemberton as "what, how and that": respectively the structure and mode of operation of a mechanism, and the difference that it makes [51]. In any particular situation at a particular time, evidence may only be available for one or the other aspect, but in nature "out there" (ontology) they are necessarily both present in all causal relationships. It is important to note that a causal relationship operates over time – the mechanism *makes* a difference – this distinguishes it from mere association, and also implies irreversibility. Making a difference implies that the outcome is distinct from what would otherwise have occurred, i.e., it involves a counterfactual.

This concept of a causal relationship has been most developed in biology. In epidemiology, difference making would correspond to the robust observation that there is an association of cigarette smoking with lung cancer, with features that strongly suggest causation. Complementary to this there must be a mechanism that explains how the association is brought about biologically. In much of biology, not only this type of example, difference making is non-quantitative: the new impulse in a post-synaptic neuron that is generated by an incoming neuronal impulse is the difference that the synaptic apparatus makes; the attack of malaria is the difference that the inoculation of *Plasmodium* organisms makes to a human [50].

The same concept applies quite generally, with appropriate modifications in different domains [50, 51]. In classical deterministic physics, difference making is expressed with extreme precision in equations, typically called "laws". An "interpretation" is not merely a *semantic* explanation that corresponds to the mathematics as is sometimes thought, but a necessary complement to it – explaining how the data/phenomena are generated [52]. Mathematics is thus only part of the description of a causal relationship, requiring a complementary mechanistic concept to complete it [53, 54]. This is clear from the fact that equations are symmetric about the equals sign, and if taken alone would imply reversibility of causation, whereas real-world causation is asymmetric. Indeed, laws expressed mathematically may not represent a causal change. For example, the Ideal Gas Law precisely relates the pressure, volume and temperature of a gas in an isolated closed container, but any causal change has to be introduced from outside either by agency or by accident. The Solar System is similarly static, despite containing objects that move, because the overall system is unchanging; the precise position of any particular planet or comet is knowable a thousand years ago or a thousand years hence.

"Mechanism" here should be broadly interpreted to mean how a causal change is brought about, not necessarily something that resembles a machine. Long before quantum theory with its attendant peculiarities, fields played a major role in physics alongside more mechanical formulations involving physical objects such as particles. Bohm was especially hostile to the idea of mechanism when applied to the micro scale [14]. The principle that difference making and mechanism are complementary merely states that alongside the mathematical expression of a causal relationship, an account is needed of the physics of *how* it happens, together with a description of the nature of the mechanism or field (etc.) that brings it about.

In physics, a mechanistic account in these terms typically corresponds to a physical meaning being given to a mathematical formulation. But often, the success of the mathematics is not matched by an equally successful account of the physics. For example, following the success of Ludwig Boltzmann's statistical mechanics model for atoms of a gas, Max Planck in 1900 adapted the mathematics to apply to radiation, but as a continuous distribution in place of Boltzmann's discontinuous one [3]. Planck realized that the probability distributions did not have the same meaning, and regarded the calculated probabilities as purely mathematical intermediaries with no good physical interpretation, a challenge subsequently taken up by Einstein [3].

More generally, providing an account of mechanism is less prominent in physics than e.g., in biology. Sometimes it is almost too obvious to mention, as when one object collides with and moves another. In other cases it is totally obscure, the classic example being Isaac Newton's law of gravitation. This clearly established the difference-making relationship (the equations), but postulated instantaneous action at a distance, which was regarded at the time as highly obscure and controversial, even by Newton himself [55]. Newton's theory is thus best regarded as *incomplete* – difference making without mechanism. When Einstein proposed that gravity is a consequence of the curvature of spacetime in his general relativity theory, he devised not only new mathematics but also a physical explanation, thereby remedying the incompleteness of Newton's gravitational theory.

This discussion is relevant to quantum physics in two ways. First, it is often thought that there are two interpretations of the wave function associated with e.g., an electron, i.e., as a material field (Schrödinger) or merely as a probability distribution (Max Born). With this conception of causation, it is not a question of choosing one or the other – on the contrary, *both are necessary*: they are respectively the mechanism and the difference-making aspects of the field, and are complementary.

Secondly, the problem with Newton's gravitational theory, instantaneous action at a distance, is related to non-locality. We have already seen that non-locality is an essential aspect of quantum phenomena, for example as manifest in entanglement. It is true that there are ways of expressing these concepts so that the non-locality of quantum theory does not entail instantaneous action at a distance [56]. But it may be preferable to be explicit that non-locality is a necessary feature of the subatomic world, as with Bohm's theory, and to find a way of re-expressing that as Einstein did for Newton.

Multiple/Partial Causation and Apparently Probabilistic Causation

The broadening of philosophy of science to include sciences other than physics has further illuminated probabilistic or stochastic causation [57]. Other developments, notably the theory of deterministic chaos, have also played a role in modifying earlier views about the inevitability of determinism [58], but do not deal with probabilistic causation itself. The central question is, what can be said *in general* about a causal relationship that is apparently probabilistic, and is analysed as such? It is possible to make some general observations that hold across the different domains, with the possible exception of quantum physics.

In the parts of biology where statistics plays a major role, such as epidemiology, there is no one-to-one relationship between cause and effect as there is in non-quantum physics (and the same applies to quantitative social science). The causal factor produces the outcome only in a proportion of cases (partial causation) so that the causal connection is merely a propensity. Coupled with this, other factors influence the same outcome (multiple causation) [42]. The two are obviously related, because each of the multiple factors must logically be insufficient to cause the outcome on its own, and conversely if a single factor is only a partial cause, this leaves space that must be filled by other causal factors.

In epidemiology, many outcomes of interest show a degree of regularity, for example the number of road deaths or cases of lung cancer in any particular country is approximately the same from one year to the next. Analysis of this phenomenon reveals that the mechanism by which it is brought about acts at the individual level, but the regularity is at the population level – and, it is impossible to know which of the individuals will be affected [59]. The particular level of the rate, or propensity, depends on the degree of causal force of the mechanism.

In the simple physical systems that were used to develop probability theory, such as tossing a coin, the outcome of a *single* coin-toss is determined by the detailed micro-level forces and spatial arrangements, but the regularity occurs at the macro level, manifest as the proportion of heads or tails in a *population* of tosses. Thus, there is a disconnect between the operation of the causal forces (micro) and the population-level regularity (macro). In probability theory it is usual to analyse simple systems and to postulate a "fair" die or coin, i.e., *assuming* equal probabilities, but in the general case of an asymmetric body we would still have regularity, in the sense of constant proportions of the different values, in a population of tosses in the long run. The micro explanation, the conjunction of forces in space that brings this about, again rests on the mechanistic aspect of the propensity.

The coin-tossing and epidemiological examples are very different, but have a similar structure – and it is found to be general across domains. The macro-level (population) outcome is manifest as a *proportion or relative frequency* after a large number of observations. The *propensity* for this to happen is derived from a mechanism acting at the micro/individual level, which carries a certain degree of causal force that differs on a case by case basis. Epidemiology itself does not investigate the mechanistic aspect, but this is done by complementary disciplines within biomedicine such as toxicology and pharmacology. These complementary aspects of probabilistic causation, association (difference making) and mechanistic propensity, apply to each partial cause, i.e., there are additional (multiple) such causes bringing about the *appearance* of randomness.

This conclusion drawn from these univariate examples holds also in the bivariate situation, of association between variables that may be causal. Such an association means that they occur together more often than they would purely by coincidence (or in the case of a protective factor, less often). Statistical analysis can be used to estimate the degree of confidence that this is true. Causal inference then depends on showing that the causation is in one direction, e.g., from smoking to cancer and not *vice versa*, nor is it due to a common causal factor or to selection. The causal mechanism again operates at the individual level, but is observable only at the population level. As in all such examples, the rate attributable to a particular exposure can be predicted but it is impossible to know which individuals will be affected.

What then is happening in the real world, in such situations that are susceptible to statistical analysis? It has proved difficult to find a satisfactory account of objective probability. As Bruno de Finetti stated at the beginning of his classic two-volume study *Theory of Probability*, "My thesis, paradoxically, and a little provocatively, but nonetheless genuinely, is simply this: PROBABILITY DOES NOT EXIST" [60]. This led to the influential interpretation of probability, that it is behavioristic or subjective, i.e., a person's degree of belief, as manifested for example in the price s/he is willing to pay for a ticket in a particular lottery.

However, if subjective probability does not merely reflect imperfect knowledge of a deterministic system, this leaves the question, what real-world phenomena does it correspond to? It would be arbitrary and useless if it did not represent some feature of reality quite accurately. This is the central question in this context, as our focus is scientific, i.e., on the natural world.

The answer can be clearly seen by focusing on the case of cigarette smoking and lung cancer: most long-term heavy smokers do not develop lung cancer, and some non-smokers get lung cancer – neither necessary nor sufficient causation is operating here. Smoking is only a *partial* cause – in any case of lung cancer *multiple* causes are operating [61]. The implication is that in general, *causes that appear to be probabilistic are actually instances of partial and multiple causation*. They can be analysed extremely well using the principles of probability and the methods of statistics, but that does not imply that the underlying causal reality is inherently probabilistic; there is no real-world analogue of probability.

Multiple causation is different from the interaction of distinct variables. For example, in the Ideal Gas Law, the pressure, volume and temperature of a gas are related in deterministic fashion. In multiple causation, by contrast, numerous variables separately influence a particular outcome. There can in addition be effect modification, in which the level of one variable affects the strength of influence of another variable [62].

Conclusion

Returning now to the quantum world, if the assumption holds that this analysis is valid in all domains including subatomic phenomena, the implication is that the apparently probabilistic features of quantum physics indicate the operation of multiple causation. There is, however, one important difference between these examples and quantum physics. Multiple causation in epidemiology and other similar disciplines leads to imprecision, whereas the unpredictability of quantum mechanics co-exists with extreme precision e.g., to eleven significant figures [63].

This indicates that "multiple" here may in fact be "dual": just one other type of causal impulse in addition to the known forces that feature in existing theory. This "other type" could not result from the properties of "ordinary" matter – photons, electrons, etc. – as these would be detected in the various experimental setups. A possibility is that entities that have hitherto been included under the heading of dark matter or dark energy interact with ordinary matter, thus generating apparently probabilistic behaviour. By bringing dark matter/energy into quantum theory, it might be possible to develop a realist account that is explicitly non-local, and that incorporates quantum phenomena including entanglement and decoherence. The nature of this interaction is currently obscure; it could for example involve multiple equilibria, which may possibly also be relevant to the question of how the electron orbits of atoms come to exist in the form that they do.

In the case of dark matter, any theory along these lines would have to allow for the fact that dark matter has mass, and it would therefore need to explain why this is not detectable in the standard experiments. In addition, the spatial distribution of dark matter is not necessarily ideal for explaining the behaviour of normal matter, as much of it exists as halos that envelop galactic disks, extending well beyond the edge of visible galaxies. This could place a large burden on the concept of non-locality.

The constraints in relation to dark energy are far less stringent, largely because so much less is known about it. Its apparently diffuse distribution means that it would be present everywhere that normal matter exists. By exploring its interaction with normal matter, it may be possible to discover more about its characteristics. In particular, a new theory of the suggested type might be able to resolve the large discrepancy between the magnitude of the cosmological constant predicted by current particle physics and its estimated empirical value.

It is doubtless too simplistic to state that dark energy "pushes" or "directs" the components of ordinary matter. But it could be that given a probabilistic tendency, dark energy combines with subatomic entities to produce a particular outcome. In the case of radioactive decay, this interaction would cause decay at a particular moment, with a population of such moments being drawn from the underlying tendency to decay as given by the exponential decay function. Similarly, in a context such as a double-slit experiment, discrete particle-like behaviour would arise from the continuous Schrödinger wave equation as a result of interaction between ordinary matter and dark energy.

References

[1] Carroll, R. (2008). *Remarks on Weyl geometry and quantum mechanics*. http://arxiv.org/pdf/0705.3921.pdf [accessed 26 April 2013]

[2] González-Díaz, P. F. Subquantum dark energy. *Phys. Rev. D* 2004, 69, 103512.

[3] Paty, M. Physical quantum states and the meaning of probability. In: *Stochastic causality*; Galavotti, M. C.,Suppes, P., Constantini, D., Eds.; CSLI Publications, Stanford, CA, 2001.

[4] Heisenberg, W. *Physics and Philosophy*; Harper and Row Publishers, Inc., New York, NY, 1962; p. 137.

[5] Mehra, J., Rechenberg, H. *The historical development of quantum theory*; Springer-Verlag New York, Inc., New York, NY, 2001.

[6] Howard, D. Who invented the Copenhagen Interpretation? A study in mythology. *Phil. Sci*. 2004, 71 (5), 669–682.

[7] Schrödinger, E. Discussion of probability relations between separated systems. *Math. Proc. Cambridge Phil. Soc*. 1935, 31 (4), 555–563. doi:10.1017/S0305004100013554[accessed 26 April 2013]

[8] Aspect, A., Grangier, P., Roger, G. Experimental Realization of Einstein-Podolsky-Rosen-Bohm Gedank enexperiment: A New Violation of Bell's Inequalities. *Phys. Rev. Lett*. 1982, 49 (2), 91-94doi:10.1103/PhysRevLett.49.91[accessed 26 April 2013]

[9] Bancal, J-D., Pironio, S., Acin, A., Liang, Y-C., Scarani, V., Gisin, N. Quantum non-locality based on finite speed causal influences leads to superluminal signalling. *Nature Phys*. 2012, 8, 867-870. doi:10.1038/nphys2460 [accessed 26 April 2013]

[10] Hartley, J. B. The spacetime approach to quantum mechanics. *Vistas Astronom*. 1993, 37, 569-583. http://arxiv.org/pdf/gr-qc/9210004v2.pdf [accessed 29 April 2013]

[11] Schlosshauer, M. (2005). *Decoherence, the measurement problem, and interpretations of quantum mechanics*. http://arxiv.org/pdf/quant-ph/0312059v4.pdf [accessed 26 April 2013]

[12] Bacciagaluppi, G. The role of decoherence in quantum mechanics. In: *The Stanford Encyclopedia of Philosophy (Winter 2003 Edition);* Zalta, E. N., Ed.http://plato.stanford.edu/archives/win2003/entries/qm-decoherence/ [accessed 26 April 2013]

[13] Bohm, D. A Suggested Interpretation of the Quantum Theory in Terms of "Hidden" Variables. 'I' and 'II', *Phys. Rev*.1952, 85, 166-179 and 180-193.

[14] Bohm, D. *Causality and Chance in Modern Physics*; Routledge and Kegan Paul: London, 1957,p. 110.

[15] Kiessling, M. K-H. Misleading signposts along the de Broglie-Bohm road to quantum mechanics. *Found Phys*. 2010, 40 (4), 418-429.

[16] Hiley, B. J. (2010). *Some remarks on the evolution of Bohm's proposals for an alternative to quantum mechanics*. http://www.bbk.ac.uk/tpru/BasilHiley/History_of_Bohm_s_QT.pdf [accessed 26 April 2013]

[17] Bell, J. S. On the Einstein Podolsky Rosen paradox. *Phys*. 1964, 1, 195-200.

[18] Valentini, A. De Broglie-Bohm pilot wave theory: many worlds in denial? In: *Many Worlds? Everett, Quantum Theory, and Reality*; Saunders, S.,

Barrett, J., Kent, A., Wallace, D., Eds.; Oxford University Press, Oxford, UK, 2010; pp. 476-509.

[19] Bohm, D., Hiley, B. J. *The undivided universe: an ontological interpretation of quantum theory*. Routledge, Abingdon, Oxon, UK, 1993.

[20] Appleby, D. M. Generic Bohmian trajectories of an isolated particle. *Found Phys*. 1999, 29, 1863-1883.

[21] Gröblacher, S., Paterek, T., Kaltenbaek, R., Brukner, Č., Żukowski, M., Aspelmeyer, M., Zeilinger, A. An experimental test of non-local realism. *Nature* 2007, 446, 871-875. doi:10.1038/nature05677 [accessed 26 April 2013]

[22] Colbeck, R., Renner, R. No extension of quantum theory can have improved predictive power. *Nature Communications* 2011, 2, 411. ArXiv:1005.5173.doi:10.1038/ncomms1416. [accessed 26 April 2013]

[23] Gisin, N. Impossibility of covariant deterministic nonlocal hidden-variable extensions of quantum theory. *Phys. Rev*. A 2011, 83, 020102.

[24] Freeman, K., McNamara, G. In: *search of dark matter*; 2006; Praxis Publishing, Chichester, UK.

[25] Raine, D., Thomas, T. *An introduction to the science of cosmology*; IOP Publishing, Bristol, UK, 2001.

[26] Koupelis, T. *In quest of the universe*. 7th edition; Jones and Bartlett Publishers, London, UK, 2011.

[27] Copi, C. J., Schramm, D. N., Turner, M. S. Big-Bang Nucleosynthesis and the Baryon Density of the Universe. *Science* 1995, 267(5195), 192–199.arXiv:astro-ph/9407006. doi:10.1126/science.7809624.PMID7809624 [accessed 26 April 2013]

[28] Bergstrom, L. Non-baryonic dark matter: Observational evidence and detection methods. *Reports Prog. Phys*. 2000, 63(5), 793–841.arXiv:hep-ph/0002126. doi:10.1088/0034-4885/63/5/2r3.[accessed 26 April 2013]

[29] Bertone, G., Hooper, D., Silk, J. Particle dark matter: Evidence, candidates and constraints. *Phys. Reports* 2005, 405(5–6), 279–390.ArXiv:hep-ph/0404175.doi:10.1016/j.physrep.2004.08.031[accessed 26 April 2013]

[30] Peter, A. H. G. *Dark matter: a brief review*. (2012) arXiv:1201.3942[astro-ph.CO].http://www.as.utexas.edu/~sj/a358-sp12/peter.darkmatter.1201.3942v1.pdf [accessed 26 April 2013]

[31] Kowalski, M., Rubin, D., Aldering, G., Agostinho, R. J., Amadon, A., Amanullah, R., et al. Improved Cosmological Constraints from New, Old and Combined Supernova Datasets. *Astrophys. J*.2008, 686(2), 749-778. arXiv:0804.4142. doi:10.1086/589937 [accessed 26 April 2013]

[32] WMAP Nine-year Results. http://map.gsfc.nasa.gov/news/ [accessed 26 April 2013]

[33] Blake, C., Brough, S., Colless, M., Contreras, C., Couch, W., Croom, S., et al. *The Wiggle Z Dark Energy Survey: the growth rate of cosmic structure since redshift* z=0.9.arXiv:1104.2948[accessed 26 April 2013]

[34] Carroll, S. The cosmological constant. *Living Rev. Relativity* 2001, 4, 1-56.

[35] Astier, P., Guy, J., Regnault, N., Pain, R., Auberg, E., Balam, D., et al. The Supernova Legacy Survey: measurement of Omega_M, Omega_Lambda and w from the first year data set. *Astronomy Astrophys*. 2006, 447, 31-48. http://arxiv.org/abs/astro-ph/0510447 [accessed 26 April 2013]

[36] Machamer, P., Darden, L., Craver, C. F. Thinking about mechanisms. *Phil. Sci.*2000, 67, 1-25.

[37] Russo, F., Williamson, J. Interpreting causality in the health sciences. *Internat. Studies Phil. Sci.*, 2007, 21, 157-170.

[38] Tabery, J. Difference mechanisms: explaining variation within mechanisms. *Biol. Philos*. 2009, 24, 645-664.

[39] Woodward, J. Causation in biology: stability, specificity, and the choice of levels of explanation. *Biol. Philos*. 2010, 25, 287-318.

[40] Russell, B. On the notion of cause. *Proc. Aristotelian Soc*. 1912-13, 13, 1-26.

[41] Norton, J. D. Is there an independent principle of causality in physics? *Brit. J. Phil. Sci*. 2009, 60, 475-486.

[42] Joffe, M. Causality and evidence discovery in epidemiology. In: *Explanation, Prediction, and Confirmation. New Trends and Old Ones Reconsidered*; Dieks, D., Gonzalez, W. J., Hartmann, S., Uebel, T., Weber, M. Eds.; Springer, Dordrecht, NL, 2011;pp 153-166.

[43] Glennan, S. Rethinking mechanistic explanations. *Phil. Sci*. 2002, 69, S342-353.

[44] Craver, F. C. *Explaining the brain: mechanisms and the mosaic unity of neuroscience*. Clarendon Press, Oxford, UK, 2007.

[45] Bechtel, W. *Mental mechanisms: philosophical perspectives on cognitive neuroscience*. Routledge, New York, NY, 2008.

[46] Hall, N. Two concepts of causation. In: *Causation and counterfactuals*; Collins, J., Hall, N., Paul, L. A., Eds.; Bradford Book/MIT Press, Cambridge, MA, 2004.

[47] Glennan, S. Singular and general causal relations: a mechanist perspective. In: *Causality in the sciences*. McKay Illari, P., Russo, F., Williamson, J., Eds.; Oxford University Press, Oxford, UK, 2011.

[48] Psillos, S. A glimpse of the secret connexion: Harmonising mechanisms with counterfactuals. *Perspect Sci.*2004, 12, 288–319.

[49] Broadbent, A. Inferring causation in epidemiology: mechanisms, black boxes, and contrasts. In: *Causality in the sciences*. McKay Illari, P., Russo, F., Williamson, J., Eds.; Oxford University Press, Oxford, UK, 2011.

[50] Joffe, M. The concept of causation in biology, and beyond. *Erkenntnis* 2013, DOI 10.1007/s10670-013-9508-6.

[51] Pemberton, J. (2011). Integrating mechanist and nomological machine ontologies to make sense of what-how-that causal evidence.http://personal.lse.ac.uk/pemberto/PNMO%20paper%20v6%200%20pdf.pdf [accessed 26 April 2013]

[52] Cox, D. R. Role of models in statistical analysis. *Stat. Sci.* 1990, 5, 169-174.

[53] Cartwright, N. *Nature's capacities and their measurement*. Oxford University Press, Oxford, 1994.

[54] Pearl, J. *Causality: models, reasoning and inference*. Cambridge University Press, New York, NY, 2000.

[55] Schliesser, E. Newtonian gravity as action at a distance. (2012) http://www.newappsblog.com/2012/07/newtonian-gravity-as-as-action-at-a-distance-you-know-a-sympathetic-process.html [accessed 26 April 2013]

[56] Berkovitz, J. 2008.Action at a distance in quantum mechanics. In: *The Stanford Encyclopedia of Philosophy* (Winter 2008 edition); Zalta, E. N., Ed.

[57] Galavotti, M. C., Suppes, P., Constantini, D. *Stochastic causality*. 2001. CSLI Publications, Stanford, CA, 2001.

[58] Prigogine, I. *The end of certainty*. The Free Press, New York, NY, 1997.

[59] Durkheim, E. *Suicide*. Routledge and Kegan Paul Ltd, London, 1952 [first published in 1897, in French].

[60] De Finetti, B. *Theory of Probability*, Vol. 1. John Wiley and Sons, New York, NY, 1974, pp. 3-4.

[61] Rothman, K. J., Greenland, S. Causation and causal inference in epidemiology. *Am. J. Public Health* 2005, 95, S144-S150.

[62] Vander Weele, T. J., Robins, J. M. Four types of effect modification: a classification based on directed acyclic graphs. *Epidemiol.* 2007, 18 (5), 561-568.

[63] Feynman, R. P. *QED: the strange theory of light and matter*. Princeton University Press, Princeton, NJ, 1985.

In: Research Development in Dark Matter Research ISBN: 978-1-62948-010-7
Editors: Nori Kinjo and Akira Nakajima

Chapter 4

SUPERHEATED LIQUIDS AND THE SEARCH FOR ASTROPARTICLE DARK MATTER

A. C. Fernandes*[1,2,*]*, T. A. Girard*[2]*, M. Felizardo*[2]*, J. G. Marques*[1,2]*,
***A. R. Ramos*[1,2]*, A. Kling*[1,2]*, T. Morlat*[3] *and J. Puibasset*[4]**
[1]Instituto Superior Técnico, Universidade Técnica de Lisboa,
Campus Tecnológico e Nuclear, Bobadela, Portugal
[2]Centro de Física Nuclear, Universidade de Lisboa, Lisboa, Portugal
[3]École Normale Supérieure de Montrouge, Montrouge, France
[4]Centre de Reserche sur la Matière Divisée-Centre National
de la Reserche Scientifique and Université d'Orléans, Orléans, France

Abstract

The use of superheated liquids in the search for astroparticle dark matter has recently begun to make a significant impact in the international dark matter search effort. Employed by only three projects worldwide (COUPP, PICASSO, SIMPLE), the advantage of the technique is that the thermodynamic operation of the devices requires radiations with linear energy transfer >100 keV μm^{-1} to induce a signal, rendering the detector effectively insensitive to the majority of traditional detector backgrounds which plague more conventional dark matter search detectors (including photons, electrons and muons). This insensitivity is not trivial, translating to a reduced data record, and simplified, more efficient analyses of the results. Acoustic detection of the resulting gas bubble permits isolation of

*E-mail address: anafer@ctn.ist.utl.pt

particle-induced events from a significantly reduced data sample, as well as nuclear recoil event discrimination. The simplicity of the detectors and instrumentation provides a relatively inexpensive and competitive search approach. We describe the salient aspects of both the technique and the three dark matter experiments which employ it, using SIMPLE as a point of reference.

1. Introduction

The direct search for weakly interacting massive particle (WIMP) dark matter is currently based on one of five techniques: scintillators, semiconductors, cryogenic bolometers, noble liquids and superheated liquids (SHL). The last, in contrast to the others, relies on the stimulated transition of a metastable liquid to its gas phase by energy absorption following particle interaction; because the transition criteria are thermodynamic, the devices can be operated at temperatures and/or pressures at which they are generally sensitive to only neutrons, alpha particles (α) and other high linear energy transfer (LET) radiations.

Only three search efforts employ the SHL technique: COUPP [1], PICASSO [2], and SIMPLE [3], using CF_3I, C_4F_{10} and C_2ClF_5 respectively. Of the three, COUPP is based on bubble chamber (BC) technology; PICASSO and SIMPLE employ superheated droplet detectors (SDDs), an emulsion of micrometric SHL droplets homogeneously suspended in a viscoelastic gel matrix. In either case, the phase transition generates a gas bubble, which can be recorded by optical or acoustic means. Both SDD projects employ acoustic, while COUPP employs both acoustic and optical (the liquid is essentially transparent, whereas the gel matrix of the SDDs is at best translucent).

The significant difference between the two approaches is that SDDs are continuously sensitive for extended periods since the overall liquid droplet population is maintained in steady-state superheated conditions despite bubble nucleation of some droplets, whereas in the BC the bulk liquid is only sensitized between nucleation events, each of which precipitates the transition of the liquid volume hence requires recompression to re-establish the metastable state and leads to measurement deadtime.

The advantage of the chamber approach is the ability to instrument large active target masses. SDDs have generally been confined to low concentration devices (< 1 wt% SHL per detector mass), for use in neutron [4-11] and heavy ion [12] detector applications, with impact in radiation dosimetry, heavy ion and cosmic ray physics, exotic particle detection and imaging in cancer therapy [13, 14]. For rare event applications such as a WIMP search, however, higher concentration detectors are required. PICASSO devices are ~ 1 wt% concentrations. SIMPLE detectors are generally of 1–2 wt%, beyond which the droplets are very close in proximity and tend to self-destruct as a result of massive sympathetic bubble nucleation and induced gel fractures.

All three SHL search programs have in 2012 produced results with impact in the dark matter search effort. Because of their fluorine content and fluorine's high spin sensitivity, they have generally contributed most to the search for spin-dependent WIMP-proton interactions; COUPP, with CF_3I, has also made a significant impact in the spin-independent sector. As seen in Fig. 1(a), the results of the three experiments tend to define the constraints on spin-dependent WIMP-proton interactions. In contrast, all remain less sensitive than the forefront experiments although COUPP is not far distant by virtue of its larger active mass (Fig. 1(b)).

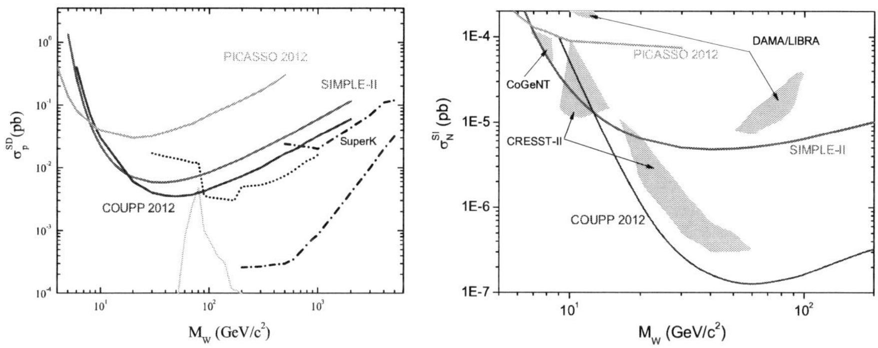

(a) (b)

Figure 1. Exclusion contours for (a) spin-dependent and (b) spin-independent WIMP-nucleus interactions of the three SHL experiments, together with leading experimental results in each sector. The region outlined in gray in (a) is favoured by cMSSM [15].

The realization of each in obtaining the results shown in Fig. 1 is different, as seen in Table 1, and largely dependent on the SHL selection. We here provide an overview of the SHL technique and its employment by each of the three experiments using Refs. [1-3] and references therein, with some focus on SIMPLE with reference to SDDs. Section 2 resumes the underlying principles of their operation. Section 3 provides descriptions of the detectors and characterizations; Section 4, the instrumentation and signal analyses. Section 5 describes the general underground installations, background estimates and data interpretations leading to the results of Fig. 1. A summary is formed in Section 6.

Table 1. Comparison between the different SHL projects

Project	SHL	Exposure (kg)	Active mass (kg)	Recoil energy threshold (keV)
COUPP [1]	CF_3I	437	3.7	7.8, 11.0, 15
PICASSO [2]	C_4F_{10}	114	0,72	1.7
SIMPLE [3]	C_2ClF_5	20.2	0.20	8.0

2. Detection Principles

2.1. Bubble Nucleation Conditions

The basic physics of a SHL detector operation, whether SDD or BC, is based on the Seitz model [16], described in detail in Refs. [17-19] and references therein. Basically, the phase transition of a SHL is triggered if the protobubble generated by the energy deposition of an incident radiation achieves a critical radius (r_c). This requires two threshold conditions for both the deposited energy (E_{dep}) and the energy deposition density, defined by the LET. The thresholds (critical energy, E_c and critical LET, LET_c) are thermodynamically defined:

$$E_{dep} \geq E_c = 4\pi r_c^2 \left(\sigma - T\frac{\partial\sigma}{\partial T}\right) + \frac{4}{3}\pi r_c^3 \rho_v h_{lv} - \frac{4}{3}\pi r_c^3 \Delta p \tag{1}$$

$$LET \geq LET_c = \frac{E_c}{\Lambda r_c} \tag{2}$$

where $r_c = 2\sigma/\Delta p$, $\sigma(T)$ is the droplet surface tension, $\Delta p = p_v - p$ is the liquid superheat, $p_v(T)$ is the vapor pressure of the SHL, p and T are the operating pressure and temperature, $h_{lv}(T)$ is the liquid-vapor heat of vaporization, and Λ is the nucleation parameter of the SHL.

Table 2 summarizes important properties of the three SHLs including the boiling temperature T_b, critical temperature T_c and critical pressure p_c.

Table 2. Comparative liquid properties of the three SHL projects at operating conditions

Project	**SHL**	T_c (°C)	T_b (°C)	ρ ($g \cdot cm^{-3}$)	r_c (μm)	h_{lv} ($kJ \cdot mole^{-1}$)	LET_c ($keV \cdot \mu m^{-1}$)
COUPP	CF_3I	123.3[20]	-21.8[20] (1 bar)	2.021	0.015	15.46[20]	62.5
PICASSO	C_4F_{10}	113.2[21]	-2.09[21] (1 bar)	1.420	0.0210	18[22]	69.8
SIMPLE	C_2ClF_5	79.95[21]	-22.2[21] (2 bar)	1.420	0.034	16[22]	176

The first term of Eq. (1) represents the work required to create the protobubble interface; the second, the energy required to evaporate the liquid during protobubble growth to r_c. The third term describes the reversible work during protobubble expansion to r_c against the liquid pressure. Generally, the second term is the largest, with the first being approximately half of the second. Not shown are various irreversible processes, e.g. sound production, which are generally small compared to the first three terms.

Generally, a nucleation event begins with a local deposition of energy in a small volume of the liquid, producing a localized, high temperature region (the "thermal spike"), whose sudden expansion produces a shock wave in the surrounding liquid. In this stage, the temperature and pressure of the liquid within the shock enclosure exceed the T_c and p_c respectively: there is no distinction between liquid and vapor, and no bubble. As the energy is transmitted from the thermalized region to the surrounding medium through shock propagation and heat conduction, the temperature and pressure of the fluid within the shock enclosure decrease, the expansion process slows and the shock wave decays. As the temperature and pressure reach T_c and p_c, a vapor-liquid interface is formed which generates a protobubble. If the deposited energy is sufficiently high, the vapor within the protobubble grows to a critical radius r_c; in case the energy was insufficient, cavity growth is impeded by interfacial and viscous forces and conduction heat loss, and the protobubble collapses.

2.2. Radiation Response

The liquid response is seen in Eq. (2) to depend on its nucleation parameter Λ, in effect defining the energy density required for bubble nucleation and yielding Λr_c as the critical effective ionic energy deposition length. A $\Lambda = \lambda\,(\rho_v/\rho_l)^{1/3}$ with $\lambda = 4.3$ has been shown in agreement with experiment for C_2ClF_5 [23] and CCl_2F_2 [24]. Because of Eq. (2), the detector response hence depends on the stopping power of the radiation within the liquid, which generally requires LET > 100 keV$\cdot\mu$m^{-1} in order to trigger bubble nucleations.

The response of various SDDs, of both low and high concentrations, to low energy neutrons, electrons, photons (γ) and cosmic-ray muons is similar [18, 25]. The threshold sensitivity to these backgrounds occurs for a reduced superheat of $s = [T\text{-}T_b]/[T_c^*\text{-}T_b] \geq 0.7$, where the $T_c^* = 0.9\,T_c$ (with all temperatures in K), since the fluid phase of organic liquids ceases to exist at a temperature about 90 % of the tabulated critical temperature T_c [26]. Operation at

$s < 0.7$ is sufficiently below threshold for these contributions to be neglected. This insensitivity is reflected in the overall data accumulation rates during the measurements. The probability for spontaneous bubble nucleation, as described by the homogeneous nucleation theory [26], is lower than 10^{-1000} nucleations per kg of SHL and day of exposure (kgd) and decreases with decreasing temperature. At $s = 0.7$, given the purity and smooth droplet/gel interfaces of the SDDs, the inhomogeneous contribution is entirely negligible.

Figure 2 displays the stopping powers for each of the target nuclei and α (He) in C_2ClF_5, as calculated via SRIM 2008 [27] using thermodynamic parameters from Refs. [21, 28].

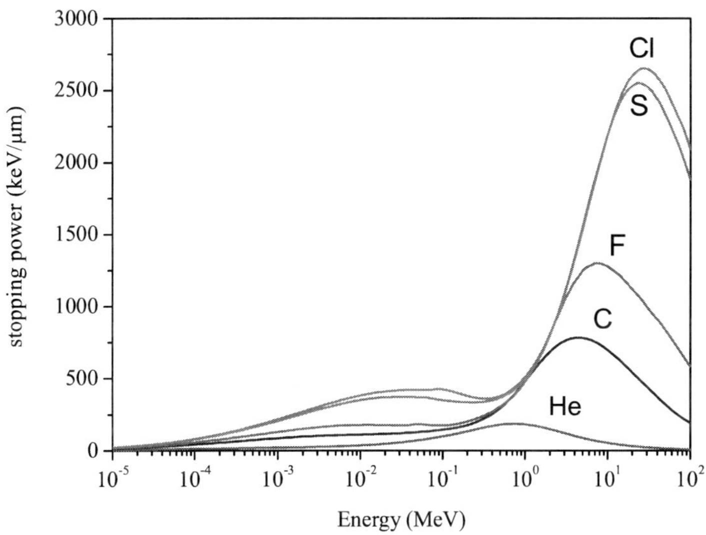

Figure 2. Stopping power of α's and recoil ions in C_2ClF_5 as a function of energy.

Penetration profiles for various ions in their respective liquids, are shown in Fig. 3 for C_2ClF_5 (at 9°C and 2 bar), C_4F_{10} (at 50°C and 1 bar), and CF_3I (at 50°C and 1 bar). Also shown are the critical LET for bubble formation at the respective temperature and pressures, obtained using $\Lambda = 4.3(\rho_l/\rho_v)^{1/3}$.

The impact of an experiment in a dark matter search depends in part on its recoil threshold energy E_{thr}, since all direct dark matter search efforts are based on the detection of nuclear recoil events generated in WIMP-nucleus elastic

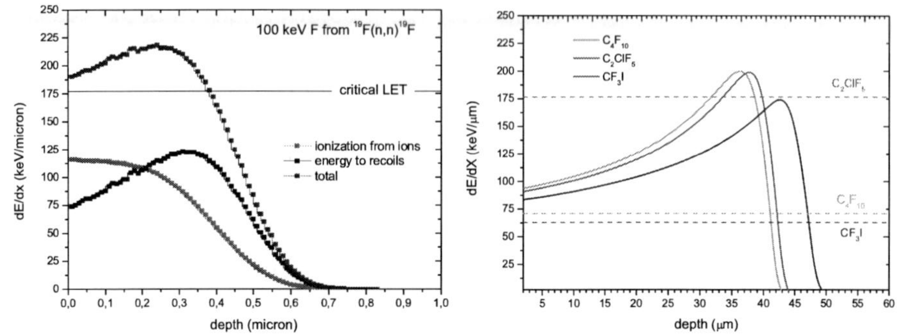

(a) (b)

Figure 3. (a) LET of recoil F ions in C_2ClF_5 as a function of penetration depth, (b) LETs of 5.5 MeV α's in the three SHLs, calculated using using $\Lambda = 4.3(\rho_l/\rho_v)^{1/3}$. The horizontal lines in each denote the corresponding critical LETs.

scatterings. The acceptance window for nuclear recoils is generically obtained from neutron calibration measurements, as elastic neutron scattering mimics the WIMP interaction. Generally, in a neutron elastic scattering on a nucleus of mass number A, the maximum ion recoil energy is given by $E^A_{recoil} = f_A E_n$, where $f_A = \frac{4A^2}{(1+A)^2}$ and E_n is the incident neutron energy. For C_2ClF_5, $f_F = 0.19$, $f_C = 0.27$ and $f_{Cl} = 0.10$ so that a detector with minimum $E_{thr} = 8$ keV implies a minimum response sensitivity to $E_n = 42$, 30 and 80 keV for F, C and Cl recoils, respectively. With ^{35}Cl, there are also two inelastic reactions which may produce events through their recoil ions: $^{35}Cl(n,p)^{35}S$ and $^{35}Cl(n,\alpha)^{32}P$. In the first case the S ion has a maximum energy of 17 keV and can produce a nucleation for neutron energies ≤ 91 keV, whereas the P ion emerges with a minimum energy of 80 keV that can always provoke a bubble. These reactions however have cross sections smaller than those of elastic scattering in Cl by $\sim$1-7 orders of magnitude, and their contribution to the detector signal is generally small (with exceptions in thermal neutron fields, e.g. at reactors).

Figure 4 shows the recoil threshold energy contours for each of the SHL constituents of the three experiments, calculated using thermodynamic parameters taken from Refs. [20, 21, 28] and $\Lambda = 4.3(\rho_l/\rho_v)^{1/3}$. The curves reflect the respective recoil ion stopping powers of Fig. 2, which for $E_{recoil} < 100$ keV are well below the respective Bragg peaks.

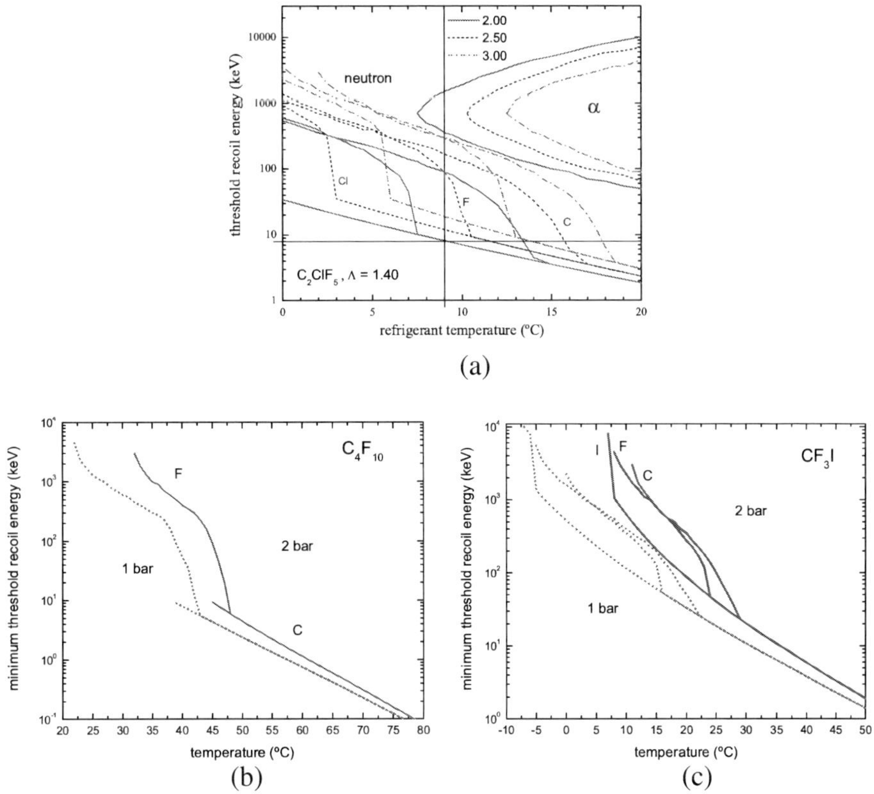

Figure 4. Temperature variations of the recoil threshold energy for each of the SHL constituents in (a) SIMPLE, (b) PICASSO, (c) COUPP; (a) includes the α threshold energy at various pressures (in bar), which is similar for (b) and (c).

Figure 4(a) also shows the similarly-calculated background α thresholds in C_2ClF_5, with similar curves occurring for C_4F_{10} and CF_3I; the region of α sensitivity lies between the two contours, which shrink concentrically and shift to higher temperatures with increasing operating pressure.

The point to note is that the different liquids require operation at different temperatures and pressures in order to achieve equally low recoil threshold energies. As the physics underlying detector operation provides only a threshold energy for bubble nucleation, an energy spectrum cannot be provided by a single measurement but rather requires measurements with a variation in temperature or pressure.

3. The Detectors

Each of the three experiments implements the SHL approach somewhat differently. For the BC approach of COUPP, the chamber used for Fig. 1 consists of a 150 mm diameter, 3 liter synthetic fused silica bell jar sealed to a flexible stainless steel bellows and immersed in propylene glycol within a stainless steel pressure vessel [1]. The propylene glycol serves as the hydraulic fluid to manage the inner pressure of the chamber, and is driven by an external pressure control unit. The bell jar contains 4.0 kg of CF_3I topped with water which isolates the CF_3I from contact with the stainless steel surfaces or seals.

For SDDs, the fabrications are somewhat more complicated by the necessity of creating an emulsion in which the liquid is homogeneously distributed throughout the emulsion with a well-defined distribution of droplet sizes; a significant difference in gel and liquid densities results in inhomogeneous distributions of differential droplet sizes within the detector. Initially, PICASSO employed a polyacrylamide, with the addition of CsCl to raise the gel density [29] which later required chemical removal because of the Cs contribution to background signal. For the Fig. 1 results however, the gel is composed of glycerin and polyethylene glycol, with the viscosity of the non-polymerized liquid used to create the homogeneous suspension [2]. Further details of the detector fabrication and purification can be found in Ref. [30]. Each detector consists of a 14 cm diameter, 40 cm high acrylic cylinder with a stainless steel cap sealed with polyurethane O-rings, containing 4.5 liter of polymerized emulsion topped by mineral oil which is connected to a hydraulic manifold in order to allow periodic pressurizations of the detectors to restore the initial superheated state. On average the active mass of a detector is 90 g $\pm$ 1 % of C_4F_{10}.

In contrast to PICASSO, SIMPLE density-matches its liquid with a 1.3 $g{\cdot}cm^{-3}$ food-based gel. The purified gel consists of biologically-clean food products, purified with respect to radioactive contaminants using actinide-specific ion-exchanging resins with the water deionized, double distilled. The gel composition is 1.71 wt% gelatin, 4.18 wt% polyvinylpyrrolidone (PVP), 15.48 wt% bi-distilled water and 78.16 wt% glycerin. The presence of the PVP (i) assists in fracture control by viscosity enhancement which decreases diffusion, (ii) improves the SDD homogeneity and reduces the droplet sizes via its surfactant behavior, (iii) decreases the liquid solubility [31], (iv) inhibits clathrate hydrate formation, and (v) reduces the migration of α-emitters to droplet boundaries as a result of actinide complex ion polarity [23]. SDDs with higher density liquids have been achieved via viscosity-matching the gel using various additives [32-34]. This approach is constrained by the gel melting at 35°C, limiting the temperature range of the device and hence the achievable E_{thr} depending on the liquid employed. Each detector consists of a 12 cm diameter glass bottle containing $\sim$ 900 g of emulsion (12-21 g of C_2ClF_5), covered by a 6 cm thick layer of glycerin and capped by a hard PVC construction containing microphone and pressure feedthroughs.

SDD construction is mitigated by the liquid solubility, which determines both the fracture probability of the gel, and the variation of the detector active mass with time. Although signals occur with or without bubble nucleation, since the liquid occupies any microscopic N_2 gas pockets formed during the fractionating stage of the suspension fabrication, the probability is increased by nucleations arising from the ambient background radiations of the fabrication site.

3.1. Detector Characterizations

3.1.1. Droplet Size Distribution

For SDDs, the distribution of droplet sizes is a crucial element in the detector response to neutrons and α's, since the recorded signals relate to the acoustic energy release in a nucleation event. The standard fabrication protocol of SIMPLE evolved from batch measurement studies to determine the droplet size distributions with respect to fractionating time and speed. Reproducible, homogeneous and normally distributed droplets of average radius $\langle r \rangle = 30 \pm 7.5\ \mu m$ as determined by optical microscopy are obtained (Fig. 5); longer fractionating times yield narrower distributions of smaller diameters; shorter times, broader distributions with larger diameter satellites.

PICASSO in contrast operates with $\langle r \rangle \sim 200\ \mu m$, as seen in Fig. 5, in order to fully accommodate the α energy loss in the droplets [29].

3.1.2. Longevity

The COUPP chamber is recompressed following each nucleation event, so that its longevity depends on the construction robustness. PICASSO, similarly, recompresses after each 40 h of operation for a period of 15 h at 6 bar [2] in order to restore the initial superheated state.

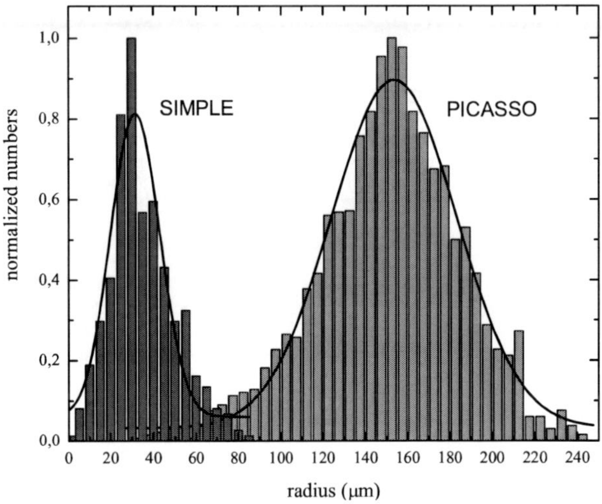

Figure 5. Typical distributions of droplet sizes in SIMPLE and PICASSO (data for PICASSO from the talk of C. Krauss at the SNOLAB 2010 Workshop).

SIMPLE runs its SDDs to exhaustion (because of its glass containment with a 4 bar tolerance, recompression is difficult). Initially, the SDDs were stable over only ~ 40 days of continuous exposure, conservatively adopted for devices without PVP in the fabrication, after which signal avalanches began to appear as a result of fractures and their propagation – which the piezoelectric instrumentation at the time was unable to discriminate from true bubble nucleations [23, 35]. The ageing process is slowed by the addition of PVP (Ref. 7 of [23]) which increases the required gel fracturing energy and viscosity, and strengthens the gel matrix while further reducing the already low solubility of C_2ClF_5. Studies made of the increased SDD lifetime and fracture reduction via use of PVP of various polymerization indices and other additives [33], were used in the development of gel recipes for long-lived SDDs based on refrigerants other than C_2ClF_5 [32]. While not directly examining the SDD longevity, these explored the rate of fracture occurrence with recipe variations employing neutron and γ irradiations.

Moreover, prior to 2006, all SIMPLE SDDs were transported overland to the utilization site in a state of reduced sensitivity ($T \leq 0°C$, $p = 4$ bar). Later analysis indicated deleterious effects on the fabrications, including fractures and bubbles (probably of mechanical origin due to transport) and the formation of clathrate hydrates [32] which provide surfaces for bubble formation when being warmed to the device operating temperature of 9°C. All science SDD fabrications since 2007 are made in an underground clean room near the measurement site, and directly installed into an experiment, yielding a loss of detection stability via fractures only after a nominal 100 days of operation. Fractures naturally resulting from Oswald ripening [36] of the bubbles ultimately lead to performance degradation – but on a significantly longer time scale.

3.1.3. Nucleation Parameter

All three experiments have measured their respective liquid Λ using a variety of techniques, to include irradiations with sources of Am/Be, ^{252}Cf and monochromatic low energy neutron beams. Measurements by SIMPLE in 1999-2000 using α-doped devices yielded $\Lambda = 1.40 \pm 0.05$ for C_2ClF_5 [23]. *In situ* measurements in 2010 [3], in which the SDD pressures were allowed to rise unchecked and the pressure at which the α-response vanished determined (see Fig. 4(a)), have confirmed this value.

PICASSO has measured the Λ of C_4F_{10} at 24°C using monochromatic low energy neutron beams [37], yielding $\Lambda = 3.8$ (higher than the $4.3(\rho_l/\rho_v)^{1/3}$ estimate of 1.13). COUPP has measured the Λ of CF_3I using Am/Be irradiations with varied temperature and pressure, yielding $\Lambda = 4$ [38].

3.1.4. Bubble Nucleation Efficiency

The bubble nucleation efficiency (ε_A) of an ion of mass number A recoiling with energy E_{recoil} is in the Seitz theory a step function at the threshold energy. The process however depends on the statistical nature of the energy deposition and its conversion into heat, which is described by [29]:

$$\varepsilon_A = 1 - \exp\left[-\Gamma\left(1 - \frac{E^A_{thr}}{E_{recoil}}\right)\right] \tag{3}$$

with Γ a detector-dependent parameter characterizing the slope of the response curve above E_{thr}, and E^A_{thr} the threshold recoil energy of the ion.

For COUPP, Eq. (3) is fit to the rates of single, double, triple and quadruple bubble events for each temperature set point and several combinations of source and source location. The free parameters are $\eta_{C,F}$ for the Seitz model and $\Gamma_{C,F}$ for Eq. (3), with the efficiency for iodine recoils fixed at 1.0 for events above threshold. Both descriptions produce acceptable fits as determined by χ^2-distribution for Poisson statistics, with the best-fit $\eta_{C,F} = 0.49 \pm 0.02$ and $\Gamma_{C,F} = 0.15 \pm 0.02$ (statistical uncertainties). Note that Eq. (3) with $\Gamma_{C,F} = 0.15$ provides a much slower rise in nucleation efficiency with energy than observed by PICASSO and SIMPLE and greatly decreases the sensitivity of the BC by cutting into the low energy portion of the recoil spectrum. Since the current COUPP data cannot distinguish between the descriptions, WIMP-nucleon interactions limits are presented as a band with edges defined by the two efficiencies; in Fig. 1, we show only the lower limit of each band, which corresponds to the Seitz model.

The rate of α pairs observed by COUPP is constant within statistical error as a function of threshold up to the cutoff, verifying that the nucleation efficiency for ^{222}Rn decays is $> 75\%$ (at 90 % C.L.) up to the ^{218}Po recoil threshold given by the Seitz model. This is consistent with results from PICASSO [37], which indicate that α-decay recoils and ^{19}F recoils in C_4F_{10} turn on sharply at the corresponding Seitz model thresholds.

For SIMPLE, analysis of monochromatic neutron irradiation data (obtained with filtered reactor neutron beams) as a function of temperature and pressure [39], yielded $\Gamma = 4.2 \pm 0.3$. Identical results were obtained for the same irradiations at 1 bar, and with different monochromaticity beams at both pressures.

For PICASSO, the principle particle-induced interactions are due to α's, since the neutron background at the underground facility is low. The Γ for these appears to depend on the α energy, with $\Gamma = 10 \pm 1$ at 146 keV and $\Gamma = 5.8 \pm 0.7$ at 71 keV (at the Bragg peak) [2]. Measurements by other authors with 17 keV mono-energetic recoils following the reaction ^{35}Cl(n,p)^{35}S are compatible with $1 < \Gamma < 5$, but with large uncertainties [40]. For the results of Fig. 1, a principal value of $\Gamma = 5$ is adopted and the parameter varied within the interval $2.5 < \Gamma < 7.5$.

4. Instrumentation & Signal Analysis

Bubble nucleations are observed by either optical or acoustic means, the latter via examination of the signal parameters. Signal analysis by COUPP encompasses both acoustic and optical channels, whereas SIMPLE and PICASSO are purely acoustic.

4.1. Instrumentation

For SIMPLE, each detector is instrumented with a pressure transducer and a single high quality electret microphone cartridge with a frequency range of 20 Hz - 16 kHz (3 dB) and a pressure sensitivity of 7.9 mV·Pa^{-1} at 1 kHz. The microphone, encased in a latex sheath, is immersed in the glycerin layer covering the emulsion. Each microphone output is preamplified by a digitally-controlled, analog microphone preamplifier characterized by low noise and harmonic distortion. The sound detection efficiency over the detector volume is 100 %. Microphone outputs are recorded in a computer via an I/O board; shielded telecommunications-grade cable linked the microphone to the connector blocks, eliminating pickup resulting from cable motion, and reducing pickup from ambient acoustic noise and site activity, even when exaggerated: the noise level is $\sim$ 2-3 mV per device [3]. The minimum voltage accuracy is 52 μV with a sensitivity of 6 μV. The DAQ acquisition rate is set at 8 kSps in a differential mode, and recorded in time-tagged files of 8 min each in a MATLAB® platform. Each pressure channel is recorded in a separate computer. Further details are reported in Ref. [41].

PICASSO employs nine piezo-transducers per detector, mounted on the outside of the acrylic container wall [2]. The transducers are ceramic disks with a diameter of 16 mm and 8.7 mm thickness, a sensitivity to 150 kHz and pressure sensitivity of 2.7 mV·Pa^{-1}. The piezoelectric sensors are read by custom-made low-noise preamplifiers; details of the DAQ system are reported in Ref. [42]. The trigger has multiplicity one: triggering of any of the nine channels causes all channels to acquire data. The trigger is fully sensitive at temperatures above 24°C (recoil energies $<$ 78 keV).

Whereas COUPP consists of a single BC, both SIMPLE and PICASSO are by nature modular; both experiments employ one "dummy" detector in the set up, not loaded with SHL but fabricated and readout as the others, to serve as a monitor for non-particle related backgrounds such as environmental noise, electronic spikes and detector cross-talk.

In COUPP, the thermodynamic state of the chamber is monitored with two temperature sensors mounted on the bellows flanges, and by pressure transducers which separately monitor the pressure of the hydraulic and inner vessel fluids. An additional fast AC-coupled pressure transducer monitors the pressure rise in the chamber to track bubble growth. Four lead zirconate piezoelectric acoustic transducers with a sensitivity to 250 kHz are attached to the exterior of the bell jar recording the acoustic emissions from bubble nucleations. The acoustic transducer signals are digitized with a 2.5 MHz sampling rate and each event is recorded for 40 ms. The signals are filtered using a single-pole high-pass filter with a cutoff at 500 Hz, and a low-pass anti-aliasing filter with a 600 kHz cutoff. Two video graphics array resolution CCD cameras are used to photograph the chamber with a 20° stereo angle at a rate of 100 frame·s^{-1}.

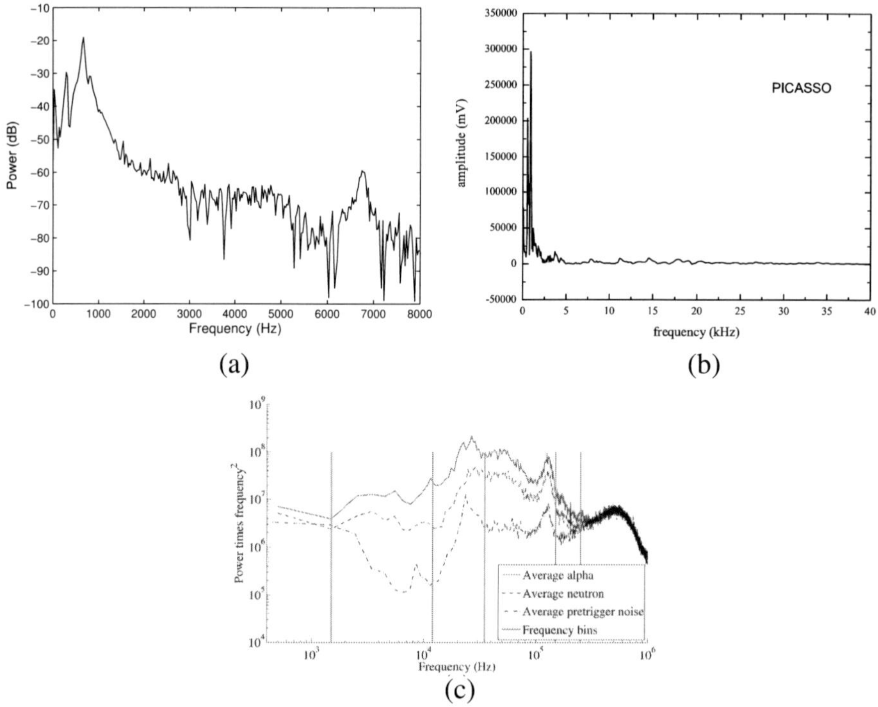

Figure 6. Typical Fast Fourier Transforms (FFT) of a particle-induced nucleation event in (a) SIMPLE, (b) PICASSO (courtesy of M-C. Piro), (c) COUPP (courtesy of A. Sonnenschein).

4.2. Data Analyses

Generally, the analysis of the acoustic signal in all three experiments proceeds in a similar fashion: (i) qualification of event registration, (ii) analysis of the corresponding temperature and pressure, and (iii) event-by-event signal analysis, in which acoustic background signals are first distinguished from those of particle-induced events, and the latter then discriminated.

In the case of SIMPLE, the pressure records are first inspected for variations of the detector sensitivity during the measurement, in order to isolate the events with pressure $\leq 2.20 \pm 0.05$ bar corresponding to an experimentally determined $E_{thr} = 8.0 \pm 0.1$ keV at 9°C. The signal records of each SDD are next inspected for raw signal rate over the measurement period, and an initial data set is formed by passing the data files through a pulse validation routine [41] which tags signal events if their amplitudes exceed the noise level of the detector by 2 mV. Tagged signals in coincidence with the dummy device are rejected, as also all candidate signals with less than five pulse spikes above threshold. The remaining set is then cross-correlated in time between all SDDs, and coincidences rejected as local noise events on the basis that a WIMP interacts with no more than one detector.

The signal waveform, decay time constant (τ), amplitude ($\mathcal{A}$), frequency (*f*) and spectral density structure (FFT) of each surviving single event are inspected individually. Following various calibrations with α's, neutrons and γ's, a particle-induced nucleation signal is characterized by a few ms time span, τ = 5-40 ms, and f between 0.45-0.75 kHz [41]. This response differs significantly from a variety of gel-associated acoustic backgrounds (such as trapped N_2 gas, gas escape, and gel fractures) which appear at lower frequencies [41], and environmental acoustic backgrounds such as human activities and experiment operation. A small sample of an exhaustive "gallery" of typical acoustic background power spectral density (psd) in SIMPLE is shown in Fig. 7.

Various correlation plots of signal parameters are constructed and used to isolate the event origin, as seen in Fig. 8.

As evident in Fig. 8(a), the totality of particle-induced events in SIMPLE appear distributed in two groups, one above $\mathcal{A}$ = 130 mV and one below $\mathcal{A}$ = 100 mV. The same effect is also observed in α and neutron irradiations, as shown in Fig. 9. Initially, these irradiations were made using separate devices for each radiation: (i) α measurements by doping 150 ml devices with a weak U_3O_8 solution following fabrication, and (ii) nuclear recoil calibrations performed on pristine SDDs using a weak Am/Be neutron source. The results of Fig. 9(a) yield a detection efficiency of > 98%. The nuclear recoil distribution is well fit by a Gaussian curve (with σ its standard deviation), which agrees well with the droplet size distribution when rescaled.

Similar calibration measurements have been performed with single SDDs,

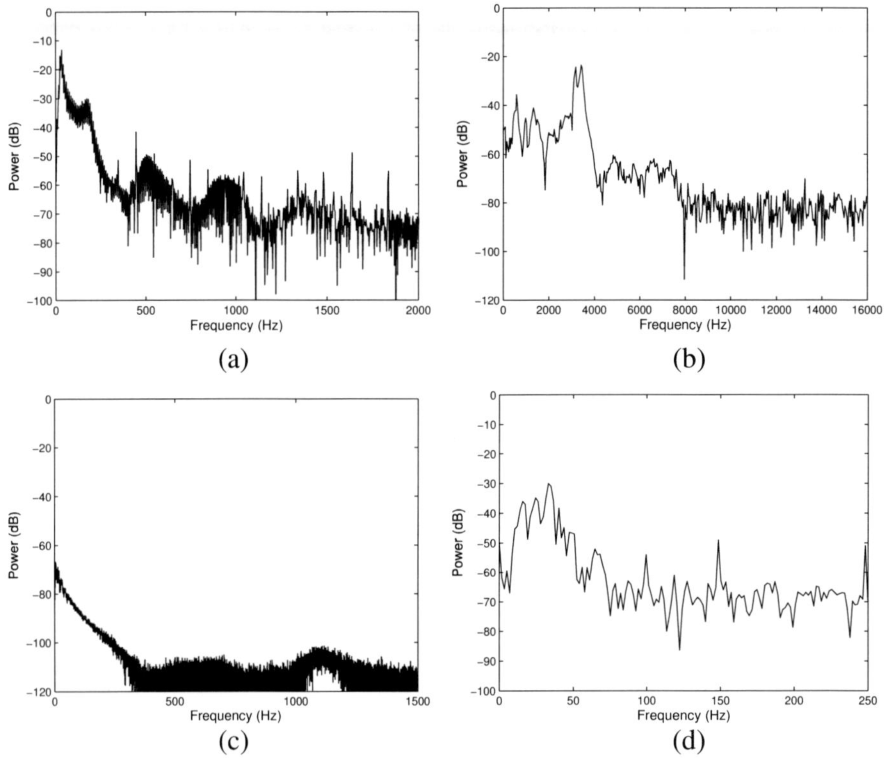

Figure 7. A partial gallery of typical gel-associated background acoustic events, indicating (a) fractures, (b) microleak, (c) pressure release and (d) N_2 release.

first with neutron irradiations and then with α-doping; a typical example, shown in Fig. 9(b), reproduces well the results of Fig. 9(a). In consequence, SIMPLE places a recoil cut of $\mathcal{A} = 100$ mV at the highest observed recoil amplitude event, i.e. at 4.4 σ above the mean of the nuclear recoil distribution of Fig. 9(a), giving an acceptance of 99.99 %. The calibration of various SDDs however yielded an upper limit range of 3.5-4.6 σ below 100 mV: rather than average, the "worst case" $\ln(A^2)$ is generally adopted as a lower limit, yielding a conservative $> 97\%$.

The analyses of PICASSO, in consequence of the higher frequency instrumentation, is based on 3 hyper-parameters [2]:

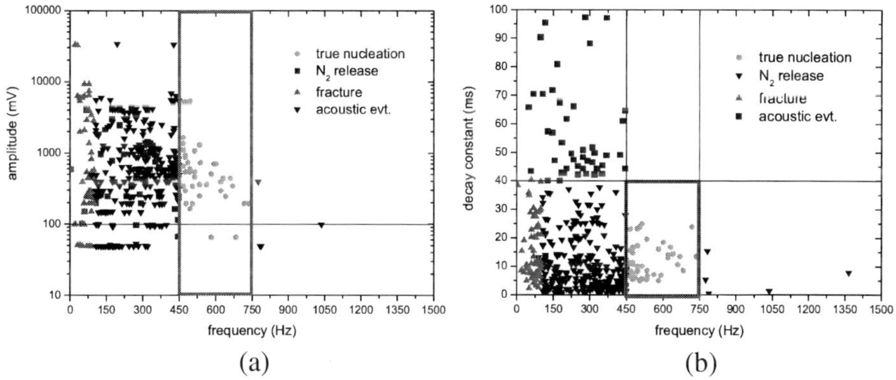

Figure 8. Scatter plots of single event signals with respect to (a) amplitudes and (b) decay constants as a function of frequency. The boxed regions correspond to particle-induced events.

- *EVAR*: the square of the waveform is integrated over the signal duration, and averaged over all active transducers to reduce solid angle effects, defining an acoustic energy threshold which is distinct from non-particle-induced noise signals. Details concerning the underlying physics processes are described in Refs. [37, 43].

- *FVAR*: since the frequency for the majority of signals is in the 20-70 kHz range, *FVAR* is defined as the logarithm of the ratio of signal power in the intervals from 20-30 kHz and 45-55 kHz. *FVAR* permits suppression of fractures or secondary events which sometimes follow a true particle-induced event, caused by a weakening of the SDD matrix and showing a significant signal power deficit in the low frequency window; environmental noise events are also efficiently removed by cuts applied on *FVAR*. A more detailed discussion of the discrimination variables and the event types which can be discriminated is found in Fig. 4 of Ref. [44].

- *RVAR*: this parameter measures the energy content within the first 25 μs after the signal start time, hence characterizes the steepness of signal growth (to identify a class of background events with a characteristic slow rise time, but an acoustic energy and frequency content comparable to particle-induced events). A scatter plot of the variables *RVAR* vs. *EVAR* at 45°C is shown in Fig. 10: particle-induced events appear in the right

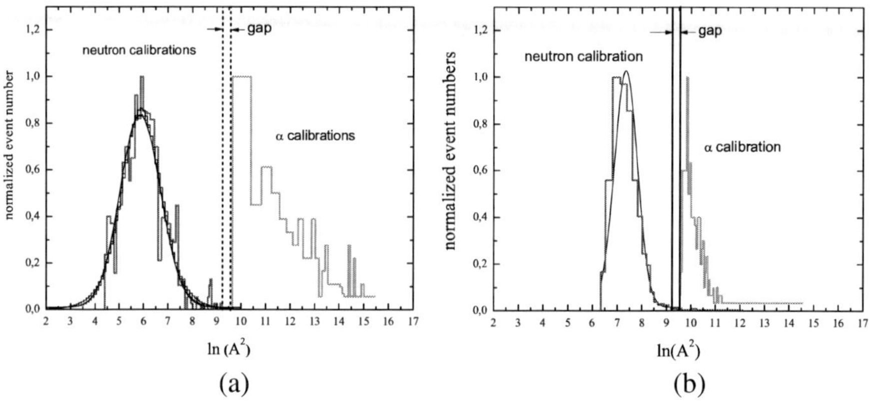

Figure 9. Histograms of calibration neutron- and α-induced nucleation events in SIMPLE SDDs: (a) separate detectors, (b) single detector with the α-doping following the neutron irradiations.

upper rectangle; secondary background events, at low values of *RVAR*.

The event qualification proceeds in the following order for each detector [2]:

1. viable runs are identified as having at least six working acoustic readout channels, with the run duration exceeding 15 h (1 h for calibrations), and the detector pressure within 0.1 bar of the ambient pressure;

2. event bursts with < 3 s (< 0.1 s for calibrations) between successive triggers are removed (the probability for such events within 3 s is negligibly small, since these are typically re-triggers of one event or induced in the detector by primary expansion);

3. an event selection on *EVAR*, using a Gaussian fit to calibration date at each temperature, with only the first 200 neutron-induced events selected in order to maintain acoustic conditions. Selection values are set to give 95 % acceptance;

4. an event selection on *RVAR*, set to yield 95 % acceptance on calibration data;

5. an event selection on *FVAR*, again to yield 95 % acceptance on calibration data.

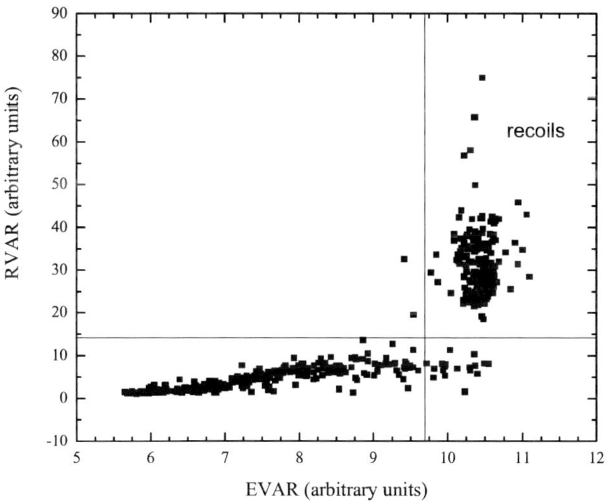

Figure 10. The signal energy *EVAR* and rise time *RVAR* variables at 45°C during calibration runs of PICASSO. Neutron-induced events cluster in the upper right rectangle with well defined *EVAR* and background events concentrated at low *RVAR* values (Reproduced from Ref. [2] with permission of the copyright owner).

The cut on *EVAR* is PICASSO's most effective discriminator of non-particle related signals; since the background increases with increasing temperature, the cut on *RVAR* becomes increasingly important at higher temperatures. The resulting event rates of all detectors are generally flat in the range of 1.7 - 92 keV (48 - 25°C), as observed via the presence of α-emitters in the droplets, and are indicative of the level of α-contamination in the individual detectors. The variation of the nucleation efficiency within range $\Gamma = 5 \pm 2.5$ results in a ± 1.5 % change in the cross section limit at 10 GeV/c^2. This uncertainty increases at lower WIMP masses and is shown as a broadening of the limits into confidence bands.

For COUPP [1], all data are subjected to a set of qualification cuts, namely that the chamber expands successfully to the desired operating pressure and is stable for more than 30 s prior to the event. Other cuts eliminate events with acoustic noise prior to the event, and events in which the video trigger failed to capture the bubble initiation. The overall efficiency for all data quality and fiducial volume cuts is 82.5 ± 1.9 %, independent of operating temperature.

Data reduction begins with examination of the photographic images to determine the number and spatial coordinates of bubbles: reconstruction of the data from two stereo views provides the spatial coordinates of the bubble to a typical accuracy of a few millimeters. The pressure rise is inspected to confirm the bubble count and identify events occurring near the vessel walls; analysis of the acoustic traces yields a characterization of event types.

The pressure rise analysis is based on data from an AC-coupled fast pressure transducer which is sampled at 10 kHz for 160 ms around the onset of a nucleation. Empirically, the rate of pressure rise is well fit by a quadratic time dependence for bubbles formed in the bulk of the target liquid, the coefficient of which is proportional to the number of bubbles in the events. The quality of the fit is uniform over the volume of the experiment except near the boundaries where it rapidly deteriorates; a modest cut on the χ^2 of the fit provides a fiducial volume definition that is uniform around the perimeter of the chamber and performs somewhat better than the stereo reconstruction of the camera images. The fiducial-to-total volume ratio is 92.1 ± 1.8 %, equivalent to removing the outer 2 mm of the liquid volume.

The pre-event baseline for each signal is examined to determine the time of bubble formation, t_0, and a FFT then constructed for $t_0 - 1$ ms $< t < t_0 + 9$ ms. The sound of a bubble nucleation exhibits a broad emission above background noise up to a frequency of 250 kHz. From calibration studies, the acoustic power is observed to vary slightly with the position of the bubble within the chamber, and the position dependence to vary with frequency. To account for these variations, the acoustic signal is separately analyzed in four frequency bands (1.5-12, 12-35, 35-150 and 150-250 kHz) which are separately corrected for spatial dependence and normalized. The acoustic event discrimination is based on a single acoustic parameter AP [45] defined by [1]:

$$\mathrm{AP} = \mathrm{a(T)} \sum_{\mathrm{j}} \mathrm{G_j} \sum_{\mathrm{n}} \mathrm{C_n}(\vec{\mathrm{x}}) \sum_{f^{\mathrm{n}}_{\mathrm{min}}}^{f^{\mathrm{n}}_{\mathrm{max}}} f \times \mathrm{psd}^{\mathrm{j}}_{\mathrm{f}} \tag{4}$$

where a(T) is an overall temperature dependent scale factor, G_j is the gain of acoustic transducer j, C_n is a correction factor for the bubble position dependence in frequency bin n, x is the position of the bubble, f is frequency, f_{min} and f_{max} are the boundaries of the frequency band, and psd^j_f is the power spectral density for the bin with center frequency f for sensor j. The AP is scaled to have a value of unity at the peak observed in its distribution for nuclear recoils induced by neutron sources. With an AP cut of $0.7 < AP < 13$, the nuclear recoil acceptance measured in calibration runs with Am/Be sources (Fig. 11) is 95.8 ± 0.5 % in the fiducial volume, resulting in a cumulative efficiency of 79.1 ± 1.9% for observing a nuclear recoil event. The efficiency for identifying multiple bubble events is stated as 100 % [1].

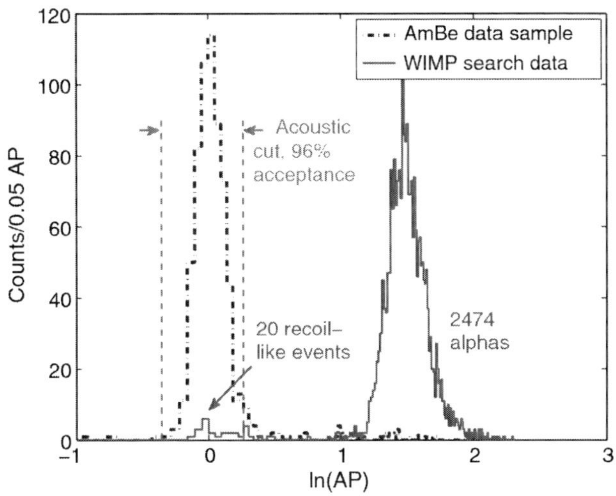

Figure 11. Data from a 553 kgd COUPP WIMP search, shown as a distribution in ln(AP) and identical analysis for data taken in the presence of an Am/Be neutron source (reproduced from Ref. [1] with permission of the copyright owner).

The difference in recoil-α event discrimination among the projects derives, at least in the case of the SDDs, from the difference in LET_c of the SHLs, and the droplet size distribution and its relationship to the α Bragg peak. The two latter provide a natural lower cutoff to the recorded α energy deposition, and are basically responsible both for the observed amplitude gap between the α and neutron populations, and also the spectral asymmetry in the α distribution. A simple explanation for these histograms derives from the fact that a droplet-neutron interaction generally results in the formation of a single protobubble, whereas an α interaction generates several. Since each protobubble constitutes an evaporation center for the droplet evaporation, the droplet evaporation rate is significantly enhanced in an α interaction, yielding an increased pressure amplitude at the microphone. This is demonstrated in Fig. 12, based on the droplet size distribution of Fig. 5, with the abscissa obtained by replacing A^2 with $r^6 n_{pb}^4$ where $n_{pb} = 12$ the number of protobubbles formed per interaction.

The decay of ^{238}U and ^{232}Th radioimpurities in the detector materials is accompanied by α's of energy 4.2-8.8 MeV, with a Bragg peak at 50 μm depth in C_2ClF_5. Since the triggering energy deposition occurs only in the Bragg peak between r = 34-42 μm (see Fig. 3(b)), droplets with dimensions smaller than 45 μm remain inert to the α passage, while those larger than 55 μm lie above the Bragg peak. This provides a natural lower cutoff to the α response distribution, which is reflected in its sharp increase at lower signal amplitudes.

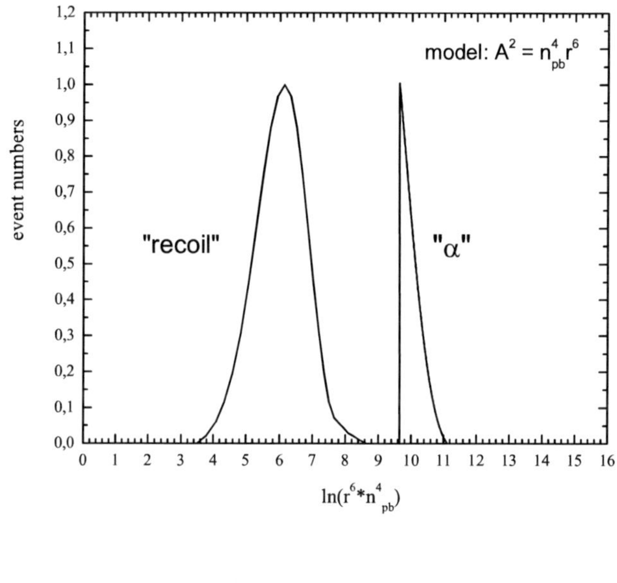

Figure 12. Recoil and α event amplitude distributions for the SIMPLE C_2ClF_5 SDD as computed via the simplified model described in the text.

5. Installation, Operation & Results Interpretations

Direct dark matter search experiments are conducted in underground laboratories. These are generally characterized by their meter water-equivalent (mwe) depth corresponding to the attenuation of muon fluence relatively to the surface.

At depths $>$ 1000 mwe, the rock overburden eliminates most neutrons of cosmogenic origin and consequently the corresponding background signal, leaving those from detector materials interior to the shielding. With SHL measurements, the main background radiations to consider are due to the U and Th decay chains present in the structural materials of the site and the set-up itself. These are both α and neutron emitters, with α's emitted by various descendants in the decay series. Neutrons are produced via spontaneous fission and by decay-induced (α,n) reactions in the U/Th-containing materials. Another aspect of concern is atmospheric radon (^{222}Rn from the U decay series) emanating both from materials in the site and externally (transport by air flow, water infiltration, etc).

SIMPLE is conducted in the 60 m^3 GESA cavern of the Laboratoire Souterrain à Bas Bruit (LSBB), located at a vertical depth of 505 m (1500 mwe) below the "Grand Montaigne" of the Plateau d'Albion near Apt in southern France. The cavern is shielded from the surrounding calcite rock by 30-100 cm of low grade concrete dating from 1971, internally sheathed by a 1 cm thickness of steel to provide a Faraday cage with electromagnetic fluctuations of order 60 fT$\sqrt{\text{Hz}}$.

For the results of Fig. 1, the measurement comprised two runs of 101 days operation each using sets of 15 SDDs with a total active mass of ~200 g. Each detector array is positioned in a plastic matrix panel 50 cm above the bottom of a 700 liter water pool maintained at a temperature of 9.0 ± 0.1°C. The water pool rests on a 10 cm thick slab of high density polyethylene, atop a 30 cm thick wood+paraffin pedestal resting on a 50 cm thick concrete floor; the pool was surrounded by three layers of sound and thermal insulation. A 50 cm thick outer water shielding, constructed of 20 liter bags-in-boxes and locally-available tap water, surrounded the insulated pool and pedestal, with a similarly-constructed overhead 75 cm water thickness. The cavern air is circulated at 0.2 m·s^{-1}, purging the cavern $\sim$ 10 times per day. The pool water is circulated at 25 liter·min^{-1}, with the entry water at the pool top so as to effectively replace the top 1 cm layer of the water pool each minute in an effort to reduce radon diffusion. The first run yielded 67 days of usable data with $p \leq 2.2$ bar equivalent to $E_{thr} = 8$ keV. In the second run the pressure was gradually increased to determine *in situ* the Λ, yielding 40 days of data with $p \leq 2.2$ bar.

Both PICASSO and COUPP are conducted in SNOLAB, Canada at 2701 m (6010 mwe) depth. In PICASSO, the detectors are installed in eight groups of four inside thermally and acoustically insulated chambers that serve as a temperature control with a precision of ± 0.1°C in the range from 20-50°C. The entire installation is surrounded by a 30.5 cm thick water shield, composed of

242 cardboard boxes containing 50 kg water-filled square polyethylene bags, with a $\sim 75\%$ filling factor.

Ten of the SDDs were used in obtaining the results in Fig. 1: seven were installed at the end of 2008 and belong to the most recent generation of PICASSO detectors with the lowest internal background; the remaining three belong to the previous generation, continuously taking data from June 2007. A science run typically lasts 40 h until recompression; a total of 264 science runs were analysed within this period yielding a total exposure of 114.3 kgd in the background and signal regions. Approximately every three months calibration data were taken at several temperatures with a weak Am/Be neutron source.

COUPP is surrounded by a water and polyethylene shield which moderates background neutrons at the experimental site to a negligible level; further shielding details are however unpublished. Data were accumulated during 223 days, corresponding to a total exposure of 553.0 kgd made at 39.0, 36.2 and 33.5°C and at $p = 2.1$ bar, corresponding to $E_{thr} = 7.8$, 11.0 and 15.5 keV. The total effective exposure for single recoil events given the 79.1 % detection efficiency described above was 437.4 kgd [1].

5.1. Radiation Background

Background estimates are generally conducted independently and "in the blind" relative to the signal analyses. For the neutron background, the results of radio-assays determining the U/Th content of surrounding shielding and detector construction materials, as well as their elemental compositions, are used as input to Monte-Carlo computations. The Rn contribution is addressed via measurement of its atmospheric concentration.

Material compositions for SIMPLE were obtained from chemical analyses of the rock and concrete composition, ion beam analysis of the hydrogen content of the wood and boron in the glass container. Standard compositions are used for the remaining materials. The densities of most materials are measured. The results of radio-assays for SIMPLE are shown in Table 3, together with the individual contributions to the detector signal. The presence of U/Th contaminations in the purified gel was measured at ~ 0.1 ppb by low-level α and γ spectroscopy of various samples of the production gel; rock, concrete and steel were measured by low-level γ spectroscopy; water and wood were assayed by α spectroscopy; neutron activation analysis was employed for paraffine and glass. Materials omitted are considered to contribute negligibly, given the close agreement between simulation results and observations.

Table 3. SIMPLE materials' radioassays

Material	^{232}Th ($Bq \cdot kg^{-1}$)	^{238}U ($Bq \cdot kg^{-1}$)	Signal contribution (evt/kgd)
Rock	0.16	5.0	~ 0
Concrete	7.7	10.5	2.9×10^{-3}
Steel	1.3×10^{-2}	3.6×10^{-3}	~ 0
Water	5.0×10^{-5}	3.2×10^{-3}	2.5×10^{-3}
Wood	3.0×10^{-3}	1.1×10^{-1}	6.7×10^{-5}
Glass	1.27	2.74	3.3×10^{-1}
Paraffin	< 0.41	< 0.25	8.4×10^{-5}

The site radon level of SIMPLE varies seasonally between 28-1000 $Bq \cdot m^{-3}$ as a result of water circulation in the mountain. Continuous purging of the site air (equivalent to $\sim$ 10 times per day) reduces the levels to $\sim$ 100 $Bq \cdot m^{-3}$. Diffusion of the environmental Rn into a detector is suppressed by the surrounding, circulating water which covers the detectors to just above their glycerin levels (6 cm above the detector matrix). With a diffusion length of 2.2 cm, the Rn concentration is reduced by a factor 10 at 7 cm below the water surface. The diffusion is also low because of the short (< 0.7 mm) Rn diffusion lengths of the SDD construction materials (glass, metal), with the measured U/Th contamination of the glass at a level significantly higher that of the gel. The N_2 over-pressuring of the device inhibits the advective influx of Rn-bearing water through the device capping, as well as diffusion of Rn from the walls of the glass container into the gel (via stiffening of the gel). The Rn contribution to the measurements, including its progeny, is estimated by time-dependent diffusion of the atmospheric Rn concentration through the water and material components. The overall α contribution to the measurement (including the detector contribution) varied between 3.26 ± 0.08 (stat) ± 0.76 (syst) evt/kgd and 5.72 ± 0.12 (stat) ± 0.29 (syst) evt/kgd because of the seasonal radon variation. The measured presence of U/Th contaminations in the SIMPLE gel yields an overall α-background level of < 0.5 evt/kgd.

At 1500 mwe, the muon fluence rate in the cavern is 2.0×10^{-2} $muon \cdot cm^{-2} s^{-1}$ and the ambient neutron fluence rate due to the surrounding calcite rock is well below 4×10^{-5} $neutron \cdot cm^{-2} s^{-1}$ [46]. As the depth of an experiment increases, the external neutron backgrounds decrease and the neutron products of the U/Th rock decay become predominant. All materials' radioassays are used as inputs to Monte-Carlo simulations of the expected neutron-generated detector event rate, which accounts for the full experimental geometries, and considers all neutron-C_2ClF_5 interactions inducing events; the source model includes spontaneous fission and decay-induced (α,n) reactions (see Ref. [47] and references therein). The results indicate negligible variations for concrete thicknesses ≥ 20 cm, and yield an expected neutron background of 0.33 ± 0.001 (stat) ± 0.038 (syst) evt/kgd, with the principle contribution (98 %) from detector glass and unpurified shield water.

The PICASSO experiment is installed at a depth of 2070 m in the Sudbury Neutrino Observatory. The cosmic muon fluence rate is reduced to less than 0.27 muon·$m^{-2}d^{-1}$ [48], and the ambient thermal neutron fluence rate from the rock has been measured as $4144.9 \pm 49.8 \pm 105.3$ neutron·$m^{-2}d^{-1}$. At the experiment location the fast neutron fluence rate (above 5 keV) has been estimated to be about 3000-4000 neutron·$m^{-2}d^{-1}$ [48, 49], with 90 % of the neutrons being produced by (α,n) reactions in the surrounding norite rock, and the remaining 10 % being spontaneous fission neutrons.

The materials were U/Th radioassayed, with the results used as input to Monte-Carlo simulations which included all structural materials, the geometric filling factor of the water boxes and self shielding effects due to the presence of other detectors within the shielding. The simulations were in agreement with measurements using ^{3}He counters. The simulations predict a reduction of fast neutrons from the shielding by a factor of 35, giving an estimated recoil event rate at the level of 1.1 neutron/kg ^{19}F/d for operation at 5 keV threshold energy.

The Rn emanation rate for these containers was measured to be only 15 ± 9 atom·d^{-1}. The detectors were fabricated in a clean environment and special efforts were made to purify the detector components in order to reduce α-emitters from the U/Th decay chains [29].

For COUPP, a variety of materials used in the BC construction were screened for their U, Th and ^{210}Po content, the latter an α emitter abundantly present in lead-containing materials such as the acoustic transducers. The (α,n) and spontaneous fission neutron production rate and energy spectrum for each material were calculated with the SOURCES-4C code [50] using the measured parent and respective daughter isotope concentrations as well as the total composition of the material in question as inputs, assuming natural abundances of any (α,n) target isotopes. The resulting neutron spectra were then used to describe the source term in Monte-Carlo simulations, and a bubble nucleation rate prediction generated for each material in the BC.

Of the materials considered, most are expected to contribute less than one event per year. The eight piezoelectric transducers and borosilicate glass viewports however contribute a significant background. Table 4 lists the predicted rates of single and multiple bubble events at the three operating thresholds assuming a bubble nucleation efficiency of 100 % on iodine and 49 % on carbon and fluorine. At each threshold, about 0.012 single bubble evts/kgd are predicted. The borosilicate viewports contribute 73 % of this rate, the piezoelectric transducers another 25 %, with the remainder produced by a combination of steel, epoxy and other components. These predictions are stated as having a systematic uncertainty of 25 % due to the assay uncertainties arising in the materials' assays, the propagation of the source term in the simulations, and from the quoted 18 % uncertainty [51] in the results from SOURCES.

COUPP also measures the efficiency with which γ interactions nucleate bubbles in the detector, using *in situ* sources of 100 μCi ^{60}Co and 1 mCi ^{133}Ba placed inside the water shield, and obtaining bubble nucleation efficiencies for single γ interactions from either source at $E_{thr} = 7.8$ keV of 1.4×10^{-8}. No response above background was observed at the two higher recoil thresholds, providing the limits shown in Table 4. The γ flux seen by the chamber with and without γ sources was measured by replacing the fused silica bell jar with a 1.78 kg NaI[Tl] scintillator, resulting in 3.4×10^{-5} evt/kgd. Assuming the nucleation probability to be independent of gamma interaction energy, the background from γ interactions is $\sim 1/3$ the neutron background at 7.8-keV threshold and negligible at 11.0 and 15.5 keV. The rate of β decays in the CF_3I is unknown: in a worst-case scenario of an atmospheric abundance of ^{14}C, the β-decay rate and resulting background would be 3 times that for gamma interactions.

The event numbers observed at the three different E_{thr} are shown in Table 5 along with the predicted events from the background simulation. The uncertainty on the Seitz threshold is calculated by combining the estimated systematic uncertainties on the temperature (1°C) and pressure (0.03 bar). The largest exposure was at $E_{thr} = 15.5 \pm 2.3$ keV with 394.0 kgd of total exposure. Including the 79.1% efficiency for detecting single bubble recoil events, the effective exposure was 311.4 kgd, yielding 8 single nuclear recoil events compared to a prediction of 3.5. At this threshold, 1 two-bubble event (with 100 % detection efficiency) was observed compared to a prediction of 1.0. Further discussion

Table 4. Predicted rates for background neutron events in COUPP arising from (α,n) reactions and spontaneous fission in the detector materials near the CF_3I volume and for background γ events from the measured ambient γ flux (from Ref. [1])

Nucleation threshold (keV)	Expected background (10^{-3} evt/kgd)			
	Neutrons			γ
	N_b=1	N_b=2	N_b=3	N_b=1
7.8 ± 1.1	12.71	3.65	1.10	4.74
11.0 ± 1.6	12.04	3.17	0.89	< 0.08
15.5 ± 2.3	11.15	2.66	0.67	< 0.01

can be found in Ref. [1].

Table 5. Observed (predicted) events for each data set of COUPP (from Ref. [1]). The final column lists single events which survived a 530 s time isolation cut

Nucleation threshold (keV)	Total exposure (kgd)	Observed (predicted) events			
		Neutrons			γ
		N_b=1	N_b=2	N_b=3	N_b=1
7.8 ± 1.1	70.6	6 (1.0)	0 (0.3)	0 (0.1)	2 (0.8)
11.0 ± 1.6	88.5	6 (0.8)	0 (0.3)	2 (0.1)	3 (0.7)
15.5 ± 2.3	394.0	6 (3.5)	1 (1.0)	0 (0.3)	8 (3.0)

Shorter exposures at $E_{thr} = 7.8 \pm 1.1$ and 11.0 ± 1.6 keV yielded 6 single nuclear recoil events each in 70.6 and 88.5 total kgd, respectively. Two three-bubble events were observed during the 11 keV exposure. The observed single recoil rates at lower threshold are significantly higher than the 0.7 and 0.8 events predicted by the neutron simulations at the 7.8 and 11.0 keV thresholds, respectively. This suggests an excess of single nuclear recoil events in the 7-15 keV range. At this low threshold, population of candidate nuclear recoil events however differs in several ways from what would be expected from true single nuclear recoils, as discussed in Ref. [1].

5.2. Interpretations

In general, all data interpretations proceed along the same line. Following measurement and analyses, the identified number of nuclear recoil-induced events is divided by the exposure to obtain the event rate dR/dE_R. The WIMP differential elastic scattering rate is given by:

$$\frac{dR}{dE_R} = \frac{\rho\varepsilon(E_R)}{2M_W}\frac{\sigma_A}{\mu^2}F^2(q)\int_{v_{min}}^{v_{max}}\frac{f(v)}{v}d^3v \tag{5}$$

where E_R is the recoil energy of the target nucleus, ρ is the local WIMP halo mass density, M_W is the WIMP mass, σ_A is the zero-momentum transfer cross section, μ is the reduced mass, $F^2(q)$ is the nuclear form factor given by $\sigma(q)/\sigma(0)$ with q the momentum transfer, v_{min} is the minimum incident WIMP speed required to produce a recoil of energy E_R, v_{max} is the maximum incident WIMP speed, ε is the efficiency of the detector, and f(v) is the WIMP velocity distribution function which depends on the WIMP halo model. The detector-dependent parameters are σ_A/μ^2, F, ε and v_{min}; v_{max} is simply the sum of the galactic escape velocity and the detector velocity with respect to the galaxy. Aside from the constant prefactor, the scattering rate can be factored into three terms: σ_A/μ^2, the nuclear form factor F^2 and the v_{min}-dependent velocity integral.

The WIMP-nucleus cross section σ_A is to first order a sum of spin-independent (SI) and spin-dependent (SD) contributions, $\sigma_A = \sigma_{SI} + \sigma_{SD}$, with

$$\sigma_{SD} = \frac{32}{\pi}G_F^2\mu_A^2\left[(a_p\langle S_p\rangle + a_n\langle S_n\rangle)^2\frac{J+1}{J}\right] \tag{6}$$

$$\sigma_{SI} = \frac{4}{\pi}G_F^2\mu_A^2\left[g_pZ + g_nN\right]^2 \tag{7}$$

where G_F is the Fermi constant, $g_{p,n}$ ($a_{p,n}$) are the SI (SD) WIMP couplings with the proton (neutron) respectively, μ_A is the WIMP-nuclide reduced mass, J is the total nuclear spin, and $\langle S_{p,n}\rangle$ are the ensemble target nucleus proton and neutron spins. With isospin conservation, $g_p = g_n = 1$ and $\sigma_{SI} \sim A^2$: in comparison with the Xe-based experiments for example, the heavier target result is enhanced by a factor of $(131/19)^2 = 48$. Since fluorine possesses the largest $\langle S_p\rangle$ of all nuclides in common use, SHLs have generally provided the most sensitive target for WIMP-proton SD studies, with less impact in the SI sector relative to their heavier counterparts owing to the A^2 enhancement of the WIMP-nucleus cross section.

In the case of SIMPLE, following completion of both the signal and background analyses, the two results are unblinded for comparison and compared. The results yielded non-zero events in the recoil acceptance window, but below the simulation estimates which suggests them to be of neutron origin. In consequence, the results were analyzed with a Feldman-Cousins (FC) method [52], with the neutron-induced recoil background estimate reduced by σ_{syst} in order to minimize the difference with observation; without this reduction, the overall event rate is reduced 30-50 %, which would improve the reported contour minima in the spin-dependent (SD) and spin-independent (SI) sectors as shown in Figs. 13 and 14 ("FC only").

In COUPP, twenty candidate nuclear recoil events and three multiple bubble events were observed, compared to a prediction of 5.3 single nuclear recoil events and 2.2 multiple bubble events from the backgrounds. All 20 nuclear recoil events were treated as dark matter candidates. Because a time isolation cut was not benchmarked prior to the measurements and given the systematic uncertainties in the neutron background simulations, no background subtraction was made.

For PICASSO, the measured rates [2] as a function of threshold energy are compared with predictions for interactions of WIMPs in our galactic halo with ^{19}F nuclei, in the presence of a constant α background in the detector, using the formalism described in [53] which approximates the recoil energy spectrum as an exponentially decreasing distribution. Since the response curves differ in shape from the flat α background of each detector for M_W smaller than 500 GeV/c^2, an upper bound on σ_A is obtained for each individual detector by fitting the WIMP response curve and the flat α background. For a given M_W the two fit parameters are σ_A and a scale factor describing the constant background. Combined in a weighted average, the maximum sensitivity occurs for WIMPs in the mass region $M_W \sim 10$ GeV/c^2, with $\sigma_A = -0.72 \pm 1.45 \pm 0.12$ pb (1 standard deviation; $\Gamma = 5$); the null result is converted into a limit [52] on the cross section for WIMP interactions of $\sigma_A = 2.00$ pb (90 % C.L.) for $\Gamma = 5$.

The resulting 90 % C.L. detection rates are customarily analyzed by all projects in the standard framework of Lewin and Smith [53], using the "standard halo model" (SHM) parameters of $\rho_{DM} = 0.3\ \mathrm{GeV{\cdot}cm^{-3}}$, $v_0 = 230\ \mathrm{km{\cdot}s^{-1}}$, $v_{esc} = 600\ \mathrm{km{\cdot}s^{-1}}$ and $v_E = 244\ \mathrm{km{\cdot}s^{-1}}$, as a means of normalizing comparisons between various experimental results (COUPP uses $v_{esc} = 544\ \mathrm{km{\cdot}s^{-1}}$, and the spin-dependent parameters from the compilation in Tovey et al. [54]). For the SD sector, the spin values of Strottman were used for ^{19}F ($\langle S_p \rangle = 0.441$, $\langle S_n \rangle =$ -

0.109) [55]); for ^{35}Cl and ^{37}Cl, the $\langle S_{p,n} \rangle$ were from Ref. [56], while the $\langle S_{p,n} \rangle$ were estimated for ^{13}C by using the odd group approximation. Contours are calculated in both sectors, including uncertainties in Λ and Γ which yield variations in the results of $\sim 1\%$; the contours presented are the least stringent.

5.2.1. Input Uncertainties: Fluorine Spin

Use of the Strottman spin for ^{19}F has been questioned [57] given the work of Divari et al. ($\langle S_p \rangle = 0.475$, $\langle S_n \rangle = -0.009$) [58] which in principle uses the more realistic interaction of Wildenthal. The impact of replacing Strottman with Divari in the interpretation (all else unchanged) is shown for SIMPLE in Fig. 13: as seen, the reach of the previous contour is increased from 5.75×10^{-3} pb to 4.98×10^{-3} pb at the contour minimum.

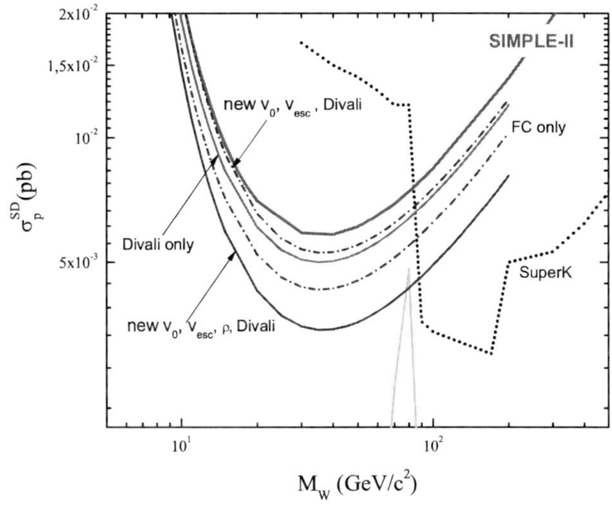

Figure 13. SIMPLE SD contours resulting from various alterations in the fluorine spin values, and the more recent SHM parameters in comparison with the contour of Fig. 1(a).

Note that both PICASSO and COUPP use the same Strottman spin values for fluorine, so that similar improvements with the Divari replacement would obtain with each, effectively renormalizing the comparison.

5.2.2. Input Uncertainties: Halo Model Parameters

Different experiments (and theoretical interpretations) use different standard halo model (SHM) parameters, comprising a somewhat confused "mix-and-match" approach with "old" and "new" SHM elements combined as indicated in Table 6. The current "canonical" values are $v_0 = 220$ km·s^{-1}, $v_{esc} = 544$ km·s^{-1} [59]. The impact of these changes is also shown ("v_0, v_{esc}") in Figs. 13 and 14, in the SD sector with the Divari spin replacement, the new contour reaches to 5.24×10^{-3} pb.

In the SI sector, for $M_W \geq 20$ GeV/c^2, there is little change; use of newer measurements of the Sun velocity relative to the Galactic halo (as high as 250 km·s^{-1} [60]) would shift the regions and the limit contours to the left, as a result of the halo variations vs. the recoil energy threshold of the measurements. Note that, in both sectors, COUPP mixes its SHM parameters, however yielding a negligible change in the reported SIMPLE contour.

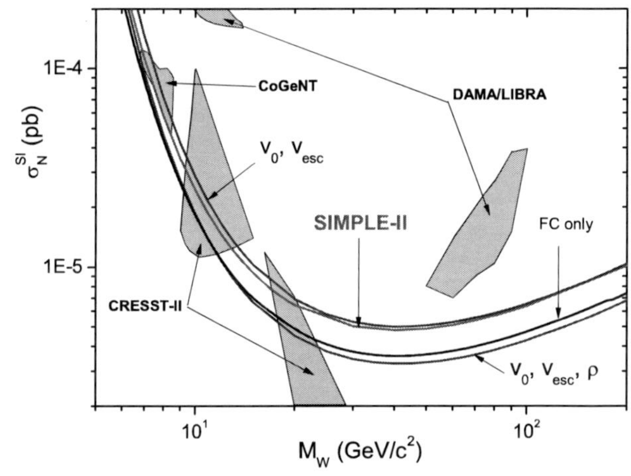

Figure 14. SIMPLE SI contours resulting from the use of more recent SHM parameters, in comparison with the contour of Fig. 1(a).

As evident in Table 6, all analyses have so far employed $\rho_{DM} = 0.3$ GeV·cm^{-3}. This parameter has also come under scrutiny in recent years, with new determinations yielding [61-64] $\rho_{DM} = 0.20$ - 0.55 GeV·cm^{-3}; new preliminary measurements from VERA suggest a revised

ρ_{DM} = 0.463 GeV·cm^{-3} [65]. The impact of a ρ_{DM} = 0.40 GeV·cm^{-3} together with the canonical v_0 and v_{esc} is also shown ("v_0,v_{esc}, ρ") in both Figs. 13 and 14. As evident, the results are strongly dependent on the SHM employed in the analyses. A complementary, halo model-independent approach has recently been suggested [66-68], based on the observation that the velocity integral of the SHM is factorable, so that the results can be recast in the form of the integral vs. v_{min}; the results must however be displayed for each M_W.

Table 6. Survey of Standard Halo Model parameters used by different reportings

Experiment	v_0 (km·s^{-1})	v_{esc} (km·s^{-1})	v_e (km·s^{-1})	ρ_{DM} (GeV·cm^{-3})
L & S	230	600	244	0.3
Canonical	220	544	244	0.3
COUPP	230	544	244	0.3
PICASSO	230	600	244	0.3
CoGeNT-08	230	650	244	0.3
XE100 (SD)	230	544	244	0.3
CDMS/EDELWEISS	220	544	232	0.3
XE100	220	544	244	0.3
CDMS	220	544	232	0.3
CRESST-II	220	544	244	0.3

6. Summary

This chapter provides an overview of the SHL technique as applied to the search for astroparticle dark matter by the PICASSO, COUPP and SIMPLE projects, each using a different liquid and implementation of the involved physics. The overall number of recorded particle-induced events reflects the general insensitivity of the approach to a majority of the common search experiment backgrounds, and an ability to reduce neutron-generated recoil backgrounds to below measurement sensitivity.

Despite the various approaches to its implementation, the SHL technique provides the basis for the most restrictive contraints on spin-dependent proton-

WIMP scattering achieved in direct dark matter searches, and contributes significantly in the spin-independent sector at low Mw via its low recoil threshold energy. As indicated, an increase in larger volume detectors with the same sensitivity yields the possibility of achieving competitivity with the noble liquid experiments. Variation of the SHL in an experiment, or between experiments, provides a further means of constraining the WIMP phase space [69].

The experimental results are however strongly dependent on both the theoretical knowledge of the target nuclei spins, and on the dark matter halo model employed in extracting the relevant cross sections. Further work in both areas, and consensus on the analyses parameters, is required.

The results of both SIMPLE and PICASSO are prejudiced by the low active mass per detector volume. This is reduced a factor 2 by replacement of the SHL with CF_3I ($\rho \sim 2$ g·cm^{-3}) as in COUPP, which is also attractive in the SI sector interpretations on account of the coherent enhancement from the I, and has previously been shown to be feasible [32]. This is not however contemplated by SIMPLE given that an $E_{thr} = 8$ keV at 2 bar corresponds to an operation temperature T $\sim$ 37°C, close to that of the gel melting.

COUPP and PICASSO have combined recently into a new experiment [70], each with the goal of achieving a 500 kg device – COUPP with a larger volume BC, and PICASSO with a "geyser" variant in which the recompression is effected non-mechanically by differential temperature and pressure variations.

SIMPLE also pursues a BC variant, originally prototyped in 2000 [23] and re-prototyped in 2010 and 2012, with a large ($\sim$ 1 kg) SHL droplet contained within a gel+glycerin-sheathed vessel; 5, 10 and 20 liter chamber designs are in progress, with the approach concept of a multi-chamber matrix containing either C_2ClF_5 or C_3F_8 in order to permit *in situ* systematic tests between the chambers. Instrumentation is the same as in the latest SIMPLE, as a result of the gel + glycerin sheathing. Inclusion of an additional 60 cm of purified water shielding immediately surrounding the detectors decreases the neutron background relative to latest setup ("Phase II") by more than three orders of magnitude.

Acknowledgments

We acknowledge and appreciate the assistance of M-C. Piro in providing information on PICASSO for the preparation of this manuscript. This work was supported by grants PDTC/FIS/115733/2009 and PTDC/FIS/121130/2010 of the Portuguese Foundation for Science and Technology (FCT), and by the Nuclear Physics Center of the University of Lisbon. The activity of M. Felizardo was supported by grant SFRH/BD/46545/2008 of FCT.

References

[1] E. Behnke et al.: *Phys. Rev.* D86 (2012) 052001.

[2] S. Archambault et al.: *Phys. Lett.* B711 (2012) 153.

[3] M. Felizardo et al.: *Phys. Rev. Lett.* 108 (2012) 201302.

[4] R. E. Apfel and S. C. Roy: *Nucl. Instrum. Meth.* 219 (1984) 582.

[5] R. E. Apfel and Y.-C. Lo: *Health Physics* 56 (1989) 79.

[6] F. d'Errico et al.: *Radiat. Prot. Dosim.* 61 (1995) 159.

[7] H. Ing et al.: *Radiat. Meas.* 27 (1997) 1.

[8] M. Das et al.: *Nucl. Instrum. Meth.* A452 (2000) 273.

[9] F. d'Errico et al.: *Nucl. Instrum. Meth.* A476 (2002) 113.

[10] F. d'Errico et al.: *Nucl. Instrum. Meth.* A476 (2002) 277.

[11] T. Sawamura et al.: *Nucl. Instrum. Meth.* A505 (2003) 29.

[12] S.-L Guo et al.: *Radiat. Meas.* 31 (1999) 167.

[13] S.-L Guo et al.: *Radiat. Meas.* 34 (2001) 269.

[14] S.-L Guo et al.: *Radiat. Meas.* 36 (2003) 183.

[15] L. Roszkowski, R. R. de Austri and R. Trotta: *J. High Energy Phys.* 07 (2007) 075.

[16] F. Seitz: *Phys. Fluids* 1 (1958) 2.

[17] M. C. Ghilea, D. D. Meyerhofer and T. C. Sangster: *Nucl. Instrum. Meth.* A648 (2011) 210.

[18] F. d'Errico: *Nucl. Instr. Meth.* B184 (2001) 229.

[19] L. K. Pan et al.: *Nucl. Instrum. Meth.* A420 (1999) 345.

[20] Y. Y. Duan et al.: *Int. J. Thermophysics* 21 (2000) 393.

[21] NIST, REPPROP code, *NIST Standard Database* 23, version 7, 2002.

[22] C. L. Yaws and M. A. Satyro: Chapter 7, Enthalpy of vaporization-Organic compounds, in: *Thermophysical Properties of Chemicals and Hydrocarbons*, ed. C. L. Yaws (William Andrew Publishing, Norwich, NY, 2009) p. 309-400.

[23] J. Puibasset: PhD thesis, Université de Paris 7, 2000 (unpublished).

[24] M. J. Harper and J. C. Rich: *Nucl. Instr. Meth.* A336 (1993) 220.

[25] M. Barnabé-Heider et al.: *Nucl. Instr. Meth.* A555 (2005) 184.

[26] J.G. Eberhart, W. Kremsner and M. Blander: *J. Colloid Interf. Sci.* 50(2) (1975) 369.

[27] http://www.srim.org/

[28] H. W. Bonin, G. R. Desnoyers and T. Cousins: *Radiat. Prot. Dosim.* 46 (2001) 265.

[29] M. Barnabé-Heider et al.: *Phys. Lett.* B624 (2005) 186.

[30] M.-C. Piro, in: *Cosmic Rays for Particle and Astroparticle Physics: Proceedings 12th ICATPP Conference*, pp. 241-249.

[31] J. D. Worley and I. M. Klotz: *Journ. Chem. Phys.* 45 (1966) 2868.

[32] T. Morlat et al.: *Astrop. Phys.* 30 (2008) 159.

[33] T. Morlat et al.: *Nucl. Instrum. Meth.* A560 (2006) 339.

[34] T. Morlat, PhD thesis, Université de Paris 7, 2003 (unpublished).

[35] J. I. Collar et al.: *New J. Phys* 2, 14.1 (2000).

[36] W. Ostwald: *Zeitschrift für Physikalische Chemie* 22 (1897) 289-330.

[37] S. Archambault et al.: *New Journ. Phys.* 13 (2011) 043006.

[38] E. Behnke et al.: *Science* 319 (2008) 933.

[39] F. Giuliani et al.: *Nucl. Instrum. Meth.* A526 (2004) 348.

[40] F. D'Errico: *Rad. Prot. Dosim.* 84 (1999) 55.

[41] M. Felizardo et al.: *Nucl. Instrum. Meth.* A589 (2008) 72.

[42] J.-P. Martin and N. Starinski: Chapter 11, *Designing Data Acquisition System for PICASSO Experiment*, in: *Data Acquisition*, ed. M. Vadursi, Sciyo, November 2010.

[43] F. Aubin et al.: *New J. Phys.* 10 (2008) 103017.

[44] S. Archambault et al.: *Phys. Lett.* B682 (2009) 185.

[45] E. Behnke et al.: *Phys. Rev. Lett.* 106 (2011) 021303.

[46] G. Waysand et al.: *Nucl. Instrum. Meth.* A444 (2000) 336.

[47] A. C. Fernandes et al.: *Nucl. Instr. Meth.* A623 (2010) 960.

[48] J. C. Loach: PhD thesis, University of Oxford, 2008.

[49] W. Davidson et al.: in: *SNO Collaboration Internal Report, Collection of Annexes in Support of Main Proposal*, 1987.

[50] W. B. Wilson et al.: *Radiat. Prot. Dosim.* 115 (2005) 117.

[51] W. S. Charlton, R. T. Perry and W. B. Wilson: *Nucl. Instrum. Meth.* B140 (1998) 1.

[52] G. J. Feldman and R. D. *Cousins: Phys. Rev.* D57 (1998) 3873.

[53] J. D. Lewin and P. F. Smith: *Astropart. Phys.* 6 (1996) 87.

[54] D. R. Tovey et al.: *Phys. Lett.* B488 (2000) 17.

[55] A. F. Pacheco and D. Strottman: *Phys. Rev.* D40 (1989) 2131.

[56] F. Giuliani and T. A. Girard: *Phys. Rev.* D71 (2005) 123503

[57] M. Cannoni: *Phys. Rev.* D84 (2011) 095017.

[58] P. C. Divari et al.: *Phys. Rev.* C61 (2000) 054612.

[59] A. Green: *Mod. Phys. Lett. A* 27 (2012) 1230004.

[60] M. J. Reid et al.: *Astrophys. J.* 700 (2009) 137.

[61] P. Salucci et al.: *Astron. Astrophys.* 523 (2010) A83.

[62] R. Catena and P. Ullio: *JCAP* 08 (2010) 004.

[63] Y. Sofue: *Publ. Astron. Soc. Japan* 64 (2012) 75.

[64] F. Iocco et al.: *JCAP* 11 (2011) 029.

[65] M. Honma: Proceedings of NDM12 — The 4th International Symposium Neutrinos and Dark Matter in Nuclear Physics, 11-15 June 2012, Nara, Japan.

[66] P. J. Fox, J. Liu and N. Weiner: *Phys. Rev.* D83 (2011) 103514.

[67] M. T. Frandsen et al.: *JCAP* 01 (2012) 024.

[68] P. Gondolo and G. B. Gelmini: *JCAP* 12 (2012) 015.

[69] G. Bertone et al.: *Phys. Rev. Lett.* 99 (2007) 151301.

[70] talks by A. Sonnenschein and T. Noble: Cosmic Frontier Workshop, SLAC, 7 March 2013.

In: Research Development in Dark Matter Research ISBN: 978-1-62948-010-7
Editors: Nori Kinjo and Akira Nakajima

Chapter 5

BEHAVIOR OF DARK MATTER IN INTERACTING DARK ENERGY MODELS

Mubasher Jamil*
Center for Advanced Mathematics and Physics (CAMP),
National University of Sciences and Technology (NUST),
Islamabad, Pakistan

Abstract

Modern cosmology has many unresolved theoretical problems dealing with the very early, present and late time evolution of the Universe. One of these unresolved and tough problems is the cosmological constant and the recent cosmic accelerated expansion detected via observations of supernova type Ia and Cosmic Microwave Background radiation. It is not very clear to cosmologists what kind of energy-matter is driving this accelerated expansion, however numerous theoretical candidates have been proposed in the literature. Given the present unexplained state of the Universe, we face a deeper uncertainty in predicting the future cosmic evolution. Cosmologists are puzzled over an enigma why the energy densities of dark matter and vacuum energy are nearly the same at the present cosmic time, so-called the Cosmic Coincidence Problem (CCP). In this Chapter, we provide a review of a phenomenological model which resolves CCP. This model involves a gravitational interaction between dark energy and dark matter in the background of a Friedmann-Robertson-Walker spacetime. The interaction ensures their energy densities are comparable at the present time irrespective of generic initial conditions of the

*E-mail address: mjamil@camp.nust.edu.pk

Universe. Besides general relativity, we investigate the interacting dark energy model in alternative gravitational theories such as loop quantum gravity, generalized teleparallel $f(T)$ gravity and Horava-Lifshitz gravity. We briefly discuss thermodynamical aspects of such a physical process.

PACS: 95.36.+x, 98.80.-k

Keywords: Dark matter; dark energy; gravitational interaction

1. Introduction

Cosmological observations obtained by SNe Ia [1], WMAP [2], SDSS [3] and X-ray [4] indicate that the observable Universe experiences an accelerated expansion. Although the simplest way to explain this behavior is the consideration of a cosmological constant [5], the known fine-tuning problem [6] led to the dark energy paradigm. The dynamical nature of dark energy, at least in an effective level, can originate from a variable cosmological "constant" [7], or from various fields, such is a canonical scalar field (quintessence) [8], a phantom field, that is a scalar field with a negative sign of the kinetic term [9], or the combination of quintessence and phantom in a unified model named quintom [10]. Finally, an interesting attempt to probe the nature of dark energy according to some basic quantum gravitational principles is the holographic dark energy paradigm [11] (although the recent developments in Horava gravity could offer a dark energy candidate with perhaps better quantum gravitational foundations [12]).

The aforementioned models offer a satisfactory description of the dark-energy behavior and its observable features. However, attributing a dynamical nature to dark energy, these scenarios introduce a new cosmological problem [6], namely why are the densities of vacuum energy and dark matter nearly equal today since these scale very differently during the expansion history. The elaboration of this "coincidence" problem led to the consideration of generalized versions of the aforementioned models with the inclusion of a coupling between dark energy and dark matter. Thus, various forms of "interacting" dark energy models have been constructed in order to fulfil the observational requirements, including interacting quintessence [13], interacting phantom [14], interacting Chaplygin gas [15] and interacting holographic dark energy [16, 17, 18]. However, one could alleviate the coincidence problem in the variable cosmological "constant" framework, too [19]. Finally, since the Universe also comprises radiation, one can consider the scenario in which dark energy interacts with both dark matter and radiation [20]. We mention that the coupling to radiation is not new, since it is also in the heart of "warm inflation" paradigm [21]. The motivation behind this extra interaction is the so-called 'triple coincidence problem' [22], that is it can offer an explanation to the fact that the radiation energy density is today only three orders of magnitude smaller than the dark-matter and dark-energy ones, although it also scales very differently. In particular, given the dark-energy-dark-matter coupling, we consider that the energy dissipated in this interaction is transferred to the radiation component and vice versa.

We adopt the following plan: In section 2, we discuss the model of interacting dark energy and dark matter in several gravitational models such as Einstein's relativistic cosmology, Horava-Lifshitz cosmology, loop quantum cosmology and the generalized teleparallel cosmology. In section 3, we analyze the generalized second law of thermodynamics in a Universe containing dark energy, dark matter and radiation. In section 4, we digress towards torsion based model to explain galactic flat rotation curves. We give a brief summary in section 5.

2. Interacting Dark Energy-Dark Matter Models

2.1. Relativistic Cosmology

We consider a homogeneous and isotropic spacetime represented by the Friedman-Robertson-Walker (FRW) line element as

$$ds^2 = -dt^2 + a(t)^2\Big[\frac{dr^2}{1-Kr^2} + r^2(d\theta^2 + \sin^2\theta d\phi^2)\Big]. \tag{1}$$

Here K describes the local curvature of spacetime. For $K = -1, 0, +1$, we have respectively spatially open, flat and closed Universes. We assume that FRW Universe is dominated by dark energy and matter (this can be dark matter, radiation and so on). The Friedman equations read (in natural units $c = G = 1$)

$$H^2 = \frac{8\pi}{3}(\rho_d + \rho_m) \tag{2}$$

$$\dot{H} = -4\pi(P_d + \rho_d + \rho_m), \tag{3}$$

where ρ_d and ρ_m are energy densities of dark energy and matter respectively. P_d is the pressure of dark energy and the Hubble parameter, $H \equiv \dot{a}/a$, is assumed to be differentiable. Although the matter part consists of both baryonic and non-baryonic (dark matter), we treat the latter component to be in gravitational interaction with dark energy. Moreover for further clarity, by taking into account the interaction, we are ruling out the purely vacuum energy component since its energy density and pressure remain fixed in the cosmic evolution for all times. Thus we hiddenly assume that the dark energy candidate is necessarily a dynamical component with the state parameter to be time dependent. Thus 'cosmological constant' (Λ) is not permissible in this scheme. In this sense, the model of interacting dark energy is an extension of ΛCDM model, in which interaction between dark energy and dark matter is forbidden.

Both matter and dark energy are allowed to exchange energy via the source term Q [23]

$$\begin{aligned} \dot{\rho}_d + 3H(w_d+1)\rho_d &= -Q \\ \dot{\rho}_m + 3H(1+w_m)\rho_m &= Q. \end{aligned} \tag{4}$$

Here subscripts d and m represent dark energy and matter terms respectively. Because of this interaction term, we do not have the conservation of partial stress-energy tensors of matter and dark energy:

$$T^{\mu\nu}_{(m);\mu} = -T^{\mu\nu}_{(d);\mu} \neq 0.$$

It is assumed that the matter component and dark energy have the same velocity which is the velocity of the whole fluid, V. We can write

$$T^{\mu\nu}_{(m);\mu} V_\nu = -T^{\mu\nu}_{(d);\mu} V_\nu = 0. \tag{5}$$

In the scalar field model of inflation, after the slow roll regime, the scalar field, whose energy density is ρ_d, decays to radiation via a rapid coherent oscillation. The source term for this decay which allows the radiation creation from inflaton is taken as [24]

$$Q = \alpha H \rho_d, \tag{6}$$

where α is a coupling constant of order unity. Also, in interacting dark energy models, different interactions between matter and dark energy are assumed and are constrained on the basis of observational data. As the nature of dark energy has not yet been known, these interactions are motivated and taken from other gravitational theories such as string theory and scalar-tensor theory and so on, or as

$$\begin{aligned} Q &= \beta H(\rho_d + \rho_m) \\ Q &= \eta H \rho_m, \end{aligned} \tag{7}$$

where β and η are real constants, are proposed phenomenologically to alleviate the coincidence problem and also to prevent the Universe to undergo the big rip [25]. To study the evolution of the Universe we are obliged to make use of a specific interaction, which we choose in a general form as [26],

$$Q = 3H(\tilde{\lambda}_m \rho_m + \lambda_d \rho_d), \tag{8}$$

where $\tilde{\lambda}_m$ and λ_d are real constants. Eq. (8) reduces to (6) and (7) for $\tilde{\lambda}_m = 0$, $\lambda_d = \tilde{\lambda}_m$ and $\lambda_d = 0$ respectively. Note that Eq. (8) is the same as the scalar

$$Q = \frac{1}{3} V_\mu V_\nu V^\gamma_{;\gamma} \left(\tilde{\lambda}_m T^{\mu\nu}_{(m)} + \lambda_d T^{\mu\nu}_{(d)} \right), \tag{9}$$

written in the comoving frame. The equation of state (EoS) parameter of dark energy, w_d, is defined by $P_d = w_d \rho_d$ and the (EoS) parameter of matter, $0 \leq w_m = \frac{P_m}{\rho_m}$, is assumed to be a constant, e.g. for cold dark matter we have $w_m = 0$ and for radiation $w_m = \frac{1}{3}$.

2.2. Horava-Lifshitz Cosmology

Concerning the gravitational background of the Universe, Hořava proposed a power-counting renormalizable theory with consistent ultra-violet (UV) behavior [27]. Although presenting an infrared (IR) fixed point, namely General Relativity, in the UV the theory exhibits an anisotropic, Lifshitz scaling between time and space. Hořava-Lifshitz gravity has been studied and extended in detail [28] and it has been applied as the cosmological framework of the Universe [29].

In this theory, the equations of motion are given by [27]

$$H^2 = \frac{\kappa^2}{6(3\Lambda - 1)}\rho_m + \frac{\kappa^2}{6(3\Lambda - 1)}\left[\frac{3\kappa^2\mu^2 k^2}{8(3\lambda - 1)a^4} + \frac{3\kappa^2\mu^2\Lambda^2}{8(3\lambda - 1)}\right] - \frac{\kappa^4\mu^2\Lambda k}{8(3\lambda - 1)^2 a^2} \tag{10}$$

$$\dot{H} + \frac{3}{2}H^2 = -\frac{\kappa^2}{4(3\Lambda - 1)} w_m \rho_m - \frac{\kappa^2}{4(3\Lambda - 1)}\left[\frac{\kappa^2\mu^2 k^2}{8(3\lambda - 1)a^4} - \frac{3\kappa^2\mu^2\Lambda^2}{8(3\lambda - 1)}\right] - \frac{\kappa^4\mu^2\Lambda k}{16(3\lambda - 1)^2 a^2}, \tag{11}$$

Here ρ_M follows the standard evolution equation

$$\dot{\rho}_m + 3H(1 + w_m)\rho_m = 0. \tag{12}$$

Observing the above Friedmann equations, concerning the dark-energy sector we can define

$$\rho_d \equiv \frac{3\kappa^2\mu^2 K^2}{8(3\lambda - 1)a^4} + \frac{3\kappa^2\mu^2\Lambda^2}{8(3\lambda - 1)} \tag{13}$$

$$p_d \equiv \frac{\kappa^2\mu^2K^2}{8(3\lambda-1)a^4} - \frac{3\kappa^2\mu^2\Lambda^2}{8(3\lambda-1)}. \tag{14}$$

The term proportional to a^{-4} is the usual "dark radiation term", present in Hořava-Lifshitz cosmology [29], while the constant term is just the explicit cosmological constant. Therefore, in expressions (13),(14) we have defined the energy density and pressure for the effective dark energy, which incorporates the aforementioned contributions. Finally, note that using (13),(14) it is straightforward to show that these dark energy quantities satisfy the standard evolution equation:

$$\dot{\rho}_d + 3H(\rho_d + p_d) = 0. \tag{15}$$

As a last step, requiring these expressions to coincide with the standard Friedmann equations, in units where $c = 1$ we set [29]:

$$G_{\text{cosmo}} = \frac{\kappa^2}{16\pi(3\lambda-1)} \tag{16}$$

$$\frac{\kappa^4\mu^2\Lambda}{8(3\lambda-1)^2} = 1, \tag{17}$$

where G_{cosmo} is the "cosmological" Newton's constant. We mention that in theories with Lorentz invariance breaking (such is Hořava-Lifshitz one) the "gravitational" Newton's constant G_{grav}, that is the one that is present in the gravitational action, does not coincide with the "cosmological" Newton's constant G_{cosmo}, that is the one that is present in Friedmann equations, unless Lorentz invariance is restored. For completeness we mention that in our case

$$G_{\text{grav}} = \frac{\kappa^2}{32\pi}. \tag{18}$$

Thus, it becomes obvious that in the IR ($\lambda = 1$), where Lorentz invariance is restored, G_{cosmo} and G_{grav} coincide.

Using the above identifications, we can re-write the Friedmann equations (10),(11) as

$$H^2 + \frac{K}{a^2} = \frac{8\pi G_{\text{cosmo}}}{3}(\rho_m + \rho_d) \tag{19}$$

$$\dot{H} + \frac{3}{2}H^2 + \frac{K}{2a^2} = -4\pi G_{\text{cosmo}}(w_m\rho_m + w_d\rho_d), \tag{20}$$

where we have introduced the effective dark energy equation-of-state parameter $w_d \equiv p_d/\rho_d$.

According to new agegraphic dark energy model [30] the dark energy sector of the Universe is attributed to a holographic dark energy, where the IR cut-of of the theory is taken to be the conformal time η of the FRW Universe:

$$\eta = \int \frac{dt}{a} = \int_0^a \frac{da}{Ha^2}. \tag{21}$$

Thus, the corresponding dark energy density reads:

$$\rho_d = \frac{3n^2}{8\pi G_{\rm grav}\eta^2}, \tag{22}$$

where the numerical factor $3n^2$ is introduced to parameterize some uncertainties, such as the species of quantum fields in the Universe, the effect of curved spacetime (since the energy density is derived for Minkowski spacetime), and so on [30].

We derive the dark energy equation-of-state parameter w_d at present as well as at small redshifts. In order to achieve this we perform the standard expansions of the literature. In particular, since $\rho_d \sim a^{-3(1+w_d)}$ we acquire [31]

$$\ln\rho_d = \ln\rho_{d0} + \frac{d\ln\rho_d}{d\ln a}\ln a + \frac{1}{2}\frac{d^2\ln\rho_d}{d(\ln a)^2}(\ln a)^2 + \dots, \tag{23}$$

where the derivatives are calculated at the present time $a_0 = 1$. Then, $w_d(\ln a)$ is given as

$$w_d(\ln a) = -1 - \frac{1}{3}\left[\frac{d\ln\rho_d}{d\ln a} + \frac{1}{2}\frac{d^2\ln\rho_d}{d(\ln a)^2}\ln a\right], \tag{24}$$

up to second order. Since

$$\rho_d = \frac{3H^2\Omega_d}{8\pi G_{\rm grav}} = \left(\frac{G_{\rm cosmo}}{G_{\rm grav}}\right)\frac{\rho_M}{\Omega_M}\Omega_d = \left(\frac{2}{3\lambda - 1}\right)\frac{\rho_{m0}a^{-3}\Omega_d}{(1-\Omega_k-\Omega_d)}, \tag{25}$$

(which holds for dust matter) the derivatives in the w_d-expansion are easily computed using the obtained expressions for Ω_d'. In addition, we can straightforwardly calculate $w_d(z)$, that is using the redshift z as the independent variable, replacing $\ln a = -\ln(1+z) \simeq -z$, which is valid for small redshifts. Doing so we obtain:

$$w_d(z) = -1 - \frac{1}{3}\left(\frac{d\ln\rho_d}{d\ln a}\right) + \frac{1}{6}\left[\frac{d^2\ln\rho_d}{d(\ln a)^2}\right] z \equiv w_0 + w_1 z. \tag{26}$$

For the interacting scenario, we have

$$w_0 = \frac{(1-\Omega_{k0})}{3n(1-\Omega_{k0}-\Omega_{d0})}\left\{n\left[\Omega_{k0} - 3 + b(\Omega_{k0}-1)\right] + \sqrt{2(3\lambda-1)}\sqrt{\Omega_{d0}}(1-\Omega_{d0}) + 3n\Omega_{d0}\right\} \tag{27}$$

$$\begin{aligned} w_1 = & \frac{(\Omega_{k0}-1)\Omega_{d0}}{12n\sqrt{3\lambda-1}\sqrt{\Omega_{d0}}(-1+\Omega_{k0}+\Omega_{d0})^2}\Big\{-\sqrt{2}n(3\lambda-1)(b-1)(\Omega_{k0}-1)^2 \\ & -2n^2\sqrt{3\lambda-1}\sqrt{\Omega_{d0}}\left[b^2(\Omega_{k0}-1)^2 - 2(\Omega_{k0}-3)\Omega_{k0} + b(\Omega_{k0}-1)(3+\Omega_{k0})\right] \\ & +2(\Omega_{k0}-1)(3\lambda-1)^{3/2}\sqrt{\Omega_{d0}} + (3\lambda-1)\Omega_{d0}^3\left(2\sqrt{3\lambda-1}\sqrt{\Omega_{d0}} - 3\sqrt{2}n\right) \\ & +(3\lambda-1)\Omega_{d0}^2\left\{6(\Omega_{k0}-1)\sqrt{3\lambda-1}\sqrt{\Omega_{d0}} - \sqrt{2}n\left[b(\Omega_{k0}-1) + 14\Omega_{k0} - 7\right]\right\} \\ & +\Omega_{d0}\Big\{\sqrt{2}n(3\lambda-1)\left\{\Omega_{k0}\left[3b(\Omega_{k0}-1) - 3\Omega_{k0} + 16\right] - 5\right\} \\ & \quad +6n^2\sqrt{3\lambda-1}\sqrt{\Omega_{d0}}\left[b(\Omega_{k0}-1) + 2\Omega_{k0}\right] - 2(3\lambda-1)^{3/2}\sqrt{\Omega_{d0}}(4\Omega_{k0}-3)\Big\}\Big\} \end{aligned} \tag{28}$$

Similarly to the previous subsection, we can insert the values and the 1σ-variation bounds for the model parameters Ω_{k0},Ω_{d0}, λ, n and the coupling parameter b, in order to extract the corresponding values for w_0 and w_1. For the first four we use $\Omega_{d0} = 0.728^{+0.015}_{-0.016}$, $\Omega_{k0} = -0.013^{+0.006}_{-0.007}$ [32], $|\lambda - 1| \leq 0.02$ [33] and $n = 2.72^{+0.11}_{-0.11}$ [34]. For the coupling parameter b there are various observational constraints in the literature [35, 36, 37], all of which lie in a narrow window around zero, thus we will use a representative interval $-0.08 < b < 0.03$ [35]. Inserting these values and variation intervals into (27),(28), within 1σ we obtain

$$\begin{aligned} w_0 &= -0.82^{+0.08}_{-0.08} \\ w_1 &= 0.08^{+0.09}_{-0.07}. \end{aligned} \tag{29}$$

As we observe, the value of w_0 is smaller than the corresponding one of the non-interacting scenario above, while w_1 is significantly larger. Furthermore, the bounds of both these parameters are significantly larger, due to the additional uncertainty in the coupling b. Finally, note moreover that these bounds are also larger from the corresponding ones in interacting new agegraphic dark energy models in conventional cosmology [34], due to the uncertainty in the running parameter λ of Hořava-Lifshitz gravity. However, there are still not so large in order to make the phantom-divide crossing possible.

2.3. Loop Quantum Cosmology

Considering quantum correction, the modified Friedmann equation turns out to be [38]

$$H^2 = \frac{\kappa^2}{3}\rho\Big(1 - \frac{\rho}{\rho_c}\Big), \tag{30}$$

where $\rho = \rho_m + \rho_d + \rho_u$, where ρ_m, ρ_d and ρ_u represent the energy densities of matter, dark energy and the radiation. Also $\rho_c \equiv \frac{\sqrt{3}}{16\pi^2\gamma^3}\rho_{\mathrm{Pl}}$ determines the loop quantum effects, ρ_{Pl} is the Planck density and γ is the dimensionless Barbero-Immirzi parameter. This parameter could be fixed as 0.2375 in order to give the area formula of black hole entropy in loop quantum gravity [39]. The observational constraints due to the quadratic term in (1) are discussed in literature [40] where it is shown that the model with quadratic correction to density is consistent with the observational tests. Thus we should not worry on the solar system tests in this model.

Due to the corrected term in (30), the big bang singularity is replaced by a quantum bounce happening at ρ_c. The bounce is supposed to happen when the matter energy density reaches the critical value ρ_c. However the numerical simulations show that modified Friedmann equations are valid in the whole cosmic evolution including the bounce [41].

Another FRW equation is

$$\dot{H} = -\frac{\kappa^2}{2}(\rho + p)\Big(1 - 2\frac{\rho}{\rho_c}\Big). \tag{31}$$

For a spatially flat Universe, the total energy conservation equation is

$$\dot{\rho} + 3H(\rho + p) = 0, \tag{32}$$

where H is the Hubble parameter, ρ is the total energy density and p is the total pressure of the background fluid.

We assume a three component fluid containing matter, dark energy and unparticle having an interaction. The corresponding continuity equations are [42]

$$\begin{aligned} \dot{\rho}_d + 3H(\rho_d + p_d) &= \Gamma_1, \\ \dot{\rho}_m + 3H\rho_m &= \Gamma_2, \\ \dot{\rho}_u + 3H(\rho_u + p_u) &= \Gamma_3, \end{aligned} \tag{33}$$

which satisfy collectively (32) such that $\Gamma_1 + \Gamma_2 + \Gamma_3 = 0$.

We define dimensionless density parameters via

$$x \equiv \frac{\kappa^2 \rho_d}{3H^2}, \quad y \equiv \frac{\kappa^2 \rho_m}{3H^2}, \quad z \equiv \frac{\kappa^2 \rho_u}{3H^2}. \tag{34}$$

Making use of parameters in (34) in the modified Friedmann equation (30), we get

$$(x+y+z)\Big(1 - \frac{3H^2}{\kappa^2}\frac{x+y+z}{\rho_c}\Big) = 1. \tag{35}$$

Using (30) and (31), we can write

$$-\frac{\dot{H}}{H^2} = \frac{3}{2}(2-x-y-z)\Big(1 + \frac{w_d x + w_u z}{x+y+z}\Big). \tag{36}$$

The equation of state parameter of the total fluid is

$$w_{\text{tot}} = \frac{p}{\rho} = \frac{w_d x + w_u z}{x+y+z}. \tag{37}$$

The continuity equations (33) in dimensionless variables reduce to

$$\begin{aligned} x' &= 3x\Big(1 + \frac{w_d x + w_u z}{x+y+z}\Big)(2-x-y-z) - 3x - 3w_d x + \frac{\kappa^2}{3H^3}\Gamma_1, \\ y' &= 3y\Big(1 + \frac{w_d x + w_u z}{x+y+z}\Big)(2-x-y-z) - 3y + \frac{\kappa^2}{3H^3}\Gamma_2, \\ z' &= 3z\Big(1 + \frac{w_d x + w_u z}{x+y+z}\Big)(2-x-y-z) - 3z - 3w_u z + \frac{\kappa^2}{3H^3}\Gamma_3. \end{aligned} \tag{38}$$

Above the primes denote differentiation with respect to $N = \ln a$. The coupling functions Γ_i, $i = 1, 2, 3$ are in general functions of the energy densities and the Hubble parameter i.e. $\Gamma_i(H\rho_i)$. The system of equations in (38) is analyzed by first equating them to zero to obtain the critical points. Next we perturb (38) up to first order about the critical points and check their stability. By choosing different forms of Γ_i, we obtain different behavior of the dynamical system. For instance, choosing $\Gamma_1 = -6bH\rho_d$, $\Gamma_2 = \Gamma_3 = 3bH\rho_d$ gives a matter dominated Universe as a stable solution, whereas $\Gamma_1 = -3bH\rho_d$, $\Gamma_2 = 3bH(\rho_d - \rho_m)$, $\Gamma_3 = 3bH\rho_m$ gives a dark energy dominated Universe as a stable solution [43].

2.4. Generalized Teleparallel Cosmology

The teleparallel gravity (TG) is an alternative of Einstein's general relativity. The TG was originally developed by Einstein in an attempt of unifying gravity and electromagnetism. The field equations for the $f(T)$ gravity are very different from those for $f(R)$ gravity, as they are second order rather than fourth order. $f(T)$ gravity is not locally Lorentz invariant and appear to harbor extra degrees of freedom not present in general relativity [44]. Here gravity is not due to curvature, but to torsion. In other word, using tetrad fields and curvature-less Weitzenbock connection instead of torsion-less Levi-Civita connection in standard general relativity.

We adopt the metric in the form of a flat Friedmann-Lemaitre-Robertson-Walker metric with metric $ds^2 = dt^2 - a(t)^2(dx^i dx_i)$, $i = 1, 2, 3$. We start with the Friedmann equation in $f(T)$ model [45]

$$H^2 = \frac{1}{1+2f_T}\left(\frac{\kappa^2}{3}\rho - \frac{f}{6}\right), \tag{39}$$

where $\rho = \rho_m + \rho_d$, and ρ_m, ρ_d represent the energy densities of matter and dark energy.

Second FRW equation is

$$\dot{H} = -\frac{\kappa^2}{2}\left(\frac{\rho + p}{1 + f_T + 2Tf_{TT}}\right). \tag{40}$$

For a spatially flat Universe ($k = 0$), the total energy conservation equation is

$$\dot{\rho} + 3H(\rho + p) = 0, \tag{41}$$

where H is the Hubble parameter, ρ is the total energy density and p is the total pressure of the background fluid. The so-called energy-balance equations corresponding to dark energy and dark matter are

$$\begin{aligned} \dot{\rho}_d + 3H(\rho_d + p_d) &= -Q, \\ \dot{\rho}_m + 3H\rho_m &= Q, \end{aligned} \tag{42}$$

We here choose the following coupling function $Q = 3Hc(\rho_d + \rho_m)$.

To perform the stability analysis of the cosmological model, it is always convenient to define dimensionless density parameters via

$$u \equiv \frac{\kappa^2 \rho_d}{3H^2}, \quad v \equiv \frac{\kappa^2 p_d}{3H^2}, \quad \Omega_m \equiv \frac{\kappa^2 \rho_m}{3H^2}. \tag{43}$$

Moreover the equation of state parameter of dark energy

$$w \equiv \frac{p_d}{\rho_d} = \frac{v}{u}. \tag{44}$$

Also the equation of state of all the fluids together is

$$w_{tot} \equiv \frac{p_d}{\rho_d + \rho_m} = \frac{v}{1 - \Omega_T}. \tag{45}$$

Using (43), we can rewrite (39) in dimensionless form

$$\Omega_m = 1 - u - \Omega_T, \tag{46}$$

where $\Omega_T \equiv \frac{f(T)}{T} - 2f_T$, is another dimensionless density parameter constructed for torsion scalar.

In this section, we treat dark energy as a perfect fluid. Physically it means that the fluid has no viscosity or heat transfer property. Its a simple fluid which cannot be self-gravitating and has a smooth distribution in space. A perfect fluid is represented by a phenomenological linear equation of state connecting energy density and pressure by

$$p_d = w\rho_d, \tag{47}$$

where w is a 'constant' of proportionality but can be a function of time or e-folding parameter $x = \ln a$. For a general dark energy paradigm, the minimum condition to be satisfied is $w \leq -1/3$. This opens the window to construct variety of theoretical models to explain cosmic acceleration. Some of these well-known models are cosmological constant ($w = -1$), quintessence ($w < -1/3$) and phantom energy ($w < -1$).

The dynamical system representing the dynamics of perfect fluid's density and pressure reads as

$$\frac{du}{dx} = -3[u + v + c(1 - \Omega_T)] + 3u\Big[\frac{1 - \Omega_T + v}{1 + 2Tf_T + f_{TT}}\Big], \tag{48}$$

$$\frac{dv}{dx} = -3w[u + v + c(1 - \Omega_T)] + 3v\Big[\frac{1 - \Omega_T + v}{1 + 2Tf_T + f_{TT}}\Big]. \tag{49}$$

We notice that our dynamical system is under-determined: three unknown functions (u, v, Ω_T) and two coupled equations. Thus we must choose $f(T)$ to solve the system. It is discussed in [46] that a power-law form of correction term $f(T) \sim T^n$ $(n > 1)$ such as T^2 can remove the finite-time future singularity. However, when $n = 0$, the correction term behaves like a cosmological constant. The model with $n = 1/2$ can be helpful in realizing power-law inflation, and also describe little-rip and pseudo-rip cosmology [46]. Due to these reasons, we choose

$$f(T) = \beta\sqrt{-T},$$

where β is a constant. Recently attractor solutions for the dynamical system with three fluids (dark matter, dark energy and radiation) interacting non-gravitationally have been investigated to resolve the coincidence problem using similar $f(T)$ [47].

To perform the stability of the system comprising equations (48) and (49), we first calculate the critical points (u_*, v_*) by equating

$$\frac{du}{dx} = 0, \ \ \frac{dv}{dx} = 0.$$

We find two critical points A and B for system (48) and (49) given in Table-1. We check the stability of the dynamical system in the neighborhood of these critical points $u = u_* + \delta u$, $v = v_* + \delta v$. This is performed by linearizing the system of equations of motion for u and v like

$$\frac{d\delta u}{dx} = \frac{3}{2}v_*\delta u + 3\left(-1 + \frac{1}{2}u_*\right)\delta v, \tag{50}$$

$$\frac{d\delta v}{dx} = -3w\delta u + 3\left(1 - w + v_*\right)\delta v, \tag{51}$$

Next we construct a Jacobian matrix from the coefficients of δu and δv in the linearized (or perturbed) system and finding the eigenvalues from it.

For the system (48) and (49), the Jacobian matrix for any critical point is

$$J = \begin{bmatrix} \frac{3}{2}v_* & -3 + \frac{3}{2}u_* \\ -3w & -3\,w + 3 + 3\,v_* \end{bmatrix}. \tag{52}$$

Table 1. Critical points and stability of the perfect fluid model ($\delta = \frac{1}{2}(w - \sqrt{w^2 + 4\,w\,c})$, $w^2 + 4\,w\,c \geq 0$)

Point	(u_*, v_*)	λ_1	λ_2	Stability Condition
A	$(2(1-c), -2)$	$3(-1 - \frac{w}{2} + \frac{1}{2}\sqrt{w^2 + 4\,w\,c})$	$3(-1 - \frac{w}{2} - \frac{1}{2}\sqrt{w^2 + 4\,w\,c})$	Unstable point
B	$(\frac{2\delta}{w}, 2\delta)$	$3(1+\delta)$	$3(\delta - w)$	Stable point if $c < \frac{w+1}{w}$

We find the eigenvalues $\lambda_{1,2}$ of the Jacobian matrix for the two critical points given in Table-1. A critical point (also called equilibrium point) is said to be *stable* if the corresponding eigenvalues are negative $\lambda_{1,2} < 0$ for all the values of the model paramcters. Such stable critical points are called attractor solutions of the dynamical system i.e. solution of the dynamical system for different initial conditions converge to stable critical point. A critical point is said to be *unstable* if both the eigenvalues are positive, and *saddle point* if one of the eigenvalues is positive. We are interested only in stable critical points. Sometimes we get *conditionally stable* critical points i.e. the critical point which can be stable only under some conditions on the model parameters. In cosmology conditionally stable points are also of interest.

In Table-2, we write down the values of relevant cosmological parameters including the total equation of state and the deceleration parameter. Note that for $f(T) = \beta\sqrt{T}$, we have $\Omega_T = 0$, therefore $w_{tot} = v_*$. For A, $q < 0$ when $c < 1/3$. We observe that A is an unstable critical point and B is conditionally stable. This means that if the interaction parameter $c < \frac{w+1}{w}$, this point is a stable critical point. Since our two dimensional phase space embedded in a three dimensional space, it has no chaotic behavior, and we can investigate the dynamical behavior of the the system, using the usual dynamical systems approach.

Table 2. Cosmological implications

Point	$w_{tot} = v_*$	$w = \frac{v_*}{u_*}$	$q = -1 - \frac{3}{2}(u_* + v_*)$
A	-2	$\frac{2}{c-1}$	$3c - 1$
B	2δ	w	$1 - 3\delta(1 + \frac{1}{w})$

3. Thermodynamics of Interacting Dark Energy Model

In this section, we introduce an interaction between dark energy, dark matter and radiation components and study the generalized second law of thermodynamics.

Specifically the energy conservation equations become [48]

$$\begin{aligned}\dot{\rho}_d + 3H(1+w_d)\rho_d &= -Q' \\ \dot{\rho}_m + 3H(1+w_m)\rho_M &= Q \\ \dot{\rho}_\chi + 3H(1+w_\chi)\rho_\chi &= Q' - Q,\end{aligned} \tag{53}$$

where a dot denotes the derivative with respect to cosmic time. In these equations we have inserted the equation-of-state parameters of the various cosmological constituents, defined as $w_i = p_i/\rho_i$, where p_i is the pressure of the corresponding constituent i. In expressions (53), Q and Q' describe the interaction terms, which can have an arbitrary form. In addition, we do not restrict the sign of these terms, that is $Q < 0$ corresponds to energy transfer from dark-matter sector to the other two constituents, $Q' > 0$ corresponds to energy transfer form dark-energy sector to the other two fluids, and $Q' < Q$ corresponds to energy loss for radiation.

The effective equation-of-state parameters are given by

$$\begin{aligned}w_d^{\text{eff}} &= w_d + \frac{Q'}{3H\rho_d} \\ w_m^{\text{eff}} &= w_m - \frac{Q}{3H\rho_m} \\ w_\chi^{\text{eff}} &= w_\chi + \frac{Q-Q'}{3H\rho_\chi}.\end{aligned} \tag{54}$$

The apparent horizon which is a marginally trapped surface with a vanishing expansion is given by

$$\tilde{r}_A = \frac{1}{\sqrt{H^2 + \frac{K}{a^2}}}. \tag{55}$$

Let us now proceed to the investigation of the generalized second law of thermodynamics in the Universe. We are going to examine whether the sum of the entropy enclosed by the apparent horizon and the entropy of the apparent horizon itself, is not a decreasing function of time. The first law of thermodynamics writes $TdS = dE + PdV$, and therefore the dark-energy and dark-matter entropies read

$$dS_i = \frac{1}{T}\Big(P_i dV + dE_i\Big), \quad i = d, m, \chi \tag{56}$$

where $V = 4\pi\tilde{r}_A^3/3$ is the volume of the system bounded by the apparent horizon and thus $dV = 4\pi\tilde{r}_A^2 d\tilde{r}_A$. From Eq. (56), we can write

$$\dot{S}_i = \frac{1}{T}\Big(P_i\, 4\pi\tilde{r}_A^2\dot{\tilde{r}}_A + \dot{E}_i\Big). \tag{57}$$

In order to connect the thermodynamically relevant quantities, namely the energies E_i and pressures P_i, with the cosmologically relevant ones, namely the energy densities ρ_i and the pressures p_i, we can straightforwardly use:

$$E_i = \frac{4\pi}{3}\tilde{r}_A^3\rho_i. \tag{58}$$

The effective equation of states are $P_i = w_i^{\text{eff}}\rho_i$. Thus (57) becomes

$$\dot{S}_i = \frac{1}{T}\left(1 + w_i^{\text{eff}}\right)\rho_i\, 4\pi\tilde{r}_A^2\left(\dot{\tilde{r}}_A - H\tilde{r}_A\right). \tag{59}$$

Using the expressions of horizon temperature $T_h = \frac{1}{2\pi\tilde{r}_A}$ and the horizon entropy $S_h = 8\pi^2\tilde{r}_A^2$, we obtain $\dot{S}_h = 16\pi^2\tilde{r}_A\dot{\tilde{r}}_A$. Thus the rate of change of total entropy becomes

$$\dot{S}_{tot} = 4\pi^2\tilde{r}_A^6 H\Big[(1+w_d)\rho_d + (1+w_m)\rho_M + (1+w_\chi)\rho_\chi\Big]^2 \geq 0. \tag{60}$$

Eq. (60) provides the expression of the generalized second law of thermodynamics in the scenario where dark energy interacts with dark matter and radiation. The fact that $\dot{S}_{tot}$ is always non-negative proves the validity of this law at all cosmological times. Moreover this result will hold independently of the interaction form, of the fluids equation-of-state parameters, and of the background geometry, provided it is FRW.

4. Explaining Flat Rotation Curves via Torsion

As is well-known that a typical spiral galaxy contains two forms of matter: luminous matter in the form of stars and stellar clusters which are found in the galactic disk while another is dark matter which is generally found in the galactic halo and encapsulates the galaxy disk. In the early Universe, the DM played crucial role in the formation of galaxies when the dense concentration of DM lied in the galactic centers which helped in the accumulation of more dust and gas to form proto-galaxies. In the later stages of galactic evolution, the DM slowly drifted towards the outer regions of the galaxies forming a huge (but less concentrated/dense) DM halos. Although the precise form of distribution of dark matter in the halos is not known, we assume that the spatial geometry of galactic halo is spherically symmetric. Moreover, the dark matter halo is isotropic: the spherical DM halo expands (hypothetically) only radially while having no tangential or orthogonal motions relative to the radial one. Jamil et al [49] investigated the problem of dark matter in spherical galaxies via torsion.

The metric of a static spherically symmetric (SSS) spacetimes can be described, without loss of generality, as

$$ds^2 = e^{a(r)}dt^2 - e^{b(r)}dr^2 - r^2\left(d\theta^2 + \sin^2\theta d\phi^2\right) \,. \tag{61}$$

This form of metric is updated by the Schwarzschild gauge, and is useful to construction of a toy model for galaxy. In order to re-write the metric (61) into the invariant form under the Lorentz transformations, we use the tetrad matrix [50]

$$\left\{e^i{}_\mu\right\} = diag\left\{e^{a(r)/2}, e^{b(r)/2}, r, r\sin\theta\right\} \,. \tag{62}$$

Using (62), one can obtain $e = \det\left[e^i{}_\mu\right] = e^{(a+b)/2}r^2\sin(\theta)$, and the torsion scalar in terms of r is given by

$$T(r) = \frac{2e^{-b}}{r}\left(a' + \frac{1}{r}\right) \tag{63}$$

where the prime ($'$) denotes the derivative with respect to the radial coordinate r. The equations of motion for an anisotropic fluid are [50]

$$4\pi\rho = \frac{f}{4} - \left(T - \frac{1}{r^2} - \frac{e^{-b}}{r}(a' + b')\right)\frac{f_T}{2}\,, \tag{64}$$

$$4\pi p_r = \left(T - \frac{1}{r^2}\right)\frac{f_T}{2} - \frac{f}{4}\,, \tag{65}$$

$$4\pi p_t = \frac{f_T}{2}\left[\frac{T}{2} + e^{-b}\left(\frac{a''}{2} + \left(\frac{a'}{4} + \frac{1}{2r}\right)(a' - b')\right)\right] - \frac{f}{4}, \tag{66}$$

$$\frac{\cot\theta}{2r^2}T' f_{TT} = 0\,, \tag{67}$$

where p_r and p_t are the radial and tangential pressures respectively, ρ is density profile.

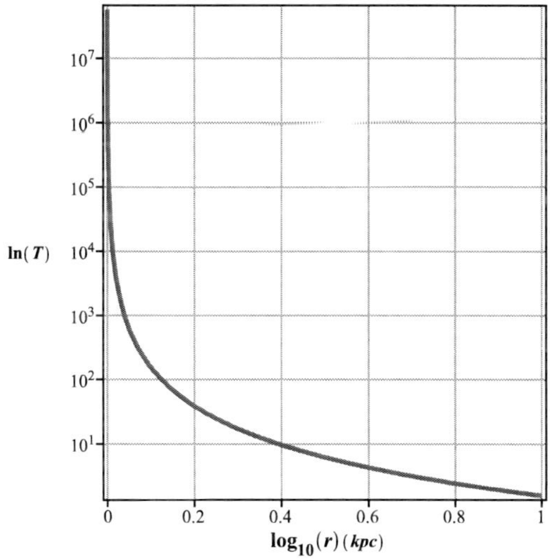

Figure 1. Non-zero scalar torsion T for metric function given by (71).

The quasi global solution for (61) with assumption $a(r) = -b(r)$ and by imposing the isotropicity in the pressure components $p_r = p_t$ is the Schwarzschild-(A)dS presented by [50]

$$e^{a(r)} = e^{-b(r)} = 1 - \frac{c_0}{r} + \frac{c_1}{3}r^2.$$

Obviously such trial metric can not be successful to generating the rotation curve of the spiral galaxies. Indeed this classical solution leads the zero torsion $T = 0$. From physical intuition we know the DM problem must arises from a non zero torsion, and specially from a variable one, $T = T(r)$. Now we introduce an ansatz for solution and choose

$$b(r) = c, \tag{68}$$

where c is an arbitrary constant. Without loss of any generalization,we assume the isotropic ansatz for the matter distribution

$$p_t = p_r. \tag{69}$$

The expression for rotation curves of galaxies is [51]

$$v_\varphi = \left(\frac{r(e^a)'}{2e^a}\right)^{1/2}, \tag{70}$$

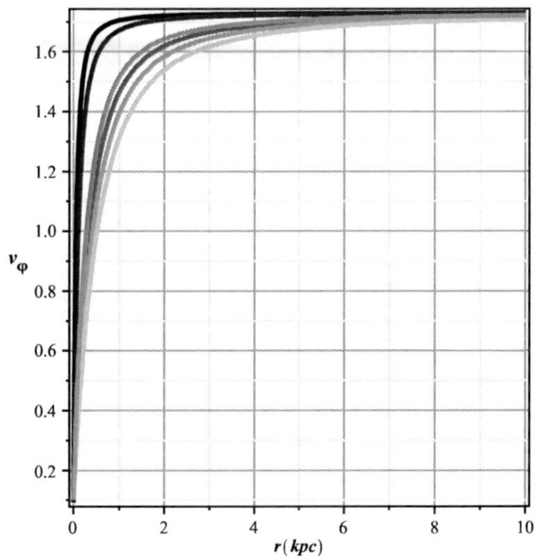

Figure 2. Toy model rotation curve for large spiral galaxy based on $f(T) \sim \sqrt{-T} + T + c_1$ model. The units of the vertical axis must be read by as factor $\times 10 \frac{Km}{s}$.

Making use of isotropic ansatz (69) in (65) and (66), we obtain

$$e^{a(r)} = \frac{r^{3-\sqrt{13-4e^c}}}{16(-13+4e^c)^2}(C_1 r^{\frac{\sqrt{13-4e^c}}{2}} - C_2)^4, \tag{71}$$

where C_1 and C_2 are new constants of integration.

Thus the flat galactic rotation curves (as shown in figures 1 & 2) can be explained in terms of torsion of space without invoking dark matter. In other words, the huge dark matter halo is nothing but mysterious and elusive torsion of space.

5. Summary

In this Chapter, we briefly reviewed one of the many approaches to alleviate the cosmic coincidence problem. The problem of having approximately same energy densities of matter and dark energy closer to our present time is nonetheless a problem of initial conditions at the formation of Universe. A Universe starting with a negligible but constant vacuum energy density will eventually become dominated by the vacuum energy. However the dynamical energy density should dominate the Universe not after the present time. A dynamical dark energy such as quintessence scalar field can decay into dark matter. In such a scenario, the late time Universe will be once again matter dominated and might collapse gravitationally into a big crunch. Any candidate of dark energy must have the observed astrophysical value at present time.

In the model of interacting dark energy and dark matter, the two dominant components of the Universe interact gravitationally. The interaction is subtle in a way to get their ratio around unity with an arbitrary initial conditions at the time of origin of Universe. The initial models of interacting dark energy-dark matter were proposed in the framework of Einstein's general relativity, which were later extended to other gravitational setups such as Horava-Lifshitz gravity, generalized teleparallel gravity and loop quantum cosmology. In all such scenarios, we like to obtain ΛCDM as an attractor solution of the dynamical equations. We reviewed such solutions above. Furthermore this scenario is also thermodynamically viable since the generalized second law of thermodynamics is valid at all times when dark energy interacts with dark matter and radiation. Finally we presented a modern approach to address the dark matter problem via torsion of spacetime and successfully obtained the flat rotation curves.

Acknowledgment

The work was supported from the research grant No. 20-2166 of the Higher Education Commission, Islamabad. The author also thank the warm hospitality of the Department of Physics, Eurasian National University, Kazakstan, during which part of this work was completed.

References

[1] A. G. Riess et al. [Supernova Search Team Collaboration], *Astron. J.* 116, 1009 (1998); S. Perlmutter et al. [Supernova Cosmology Project Collaboration], *Astrophys. J.* 517, 565 (1999).

[2] C. L. Bennett et al., *Astrophys. J. Suppl.* 148, 1 (2003).

[3] M. Tegmark et al. [SDSS Collaboration], *Phys. Rev. D* 69, 103501 (2004).

[4] S. W. Allen, et al., *Mon. Not. Roy. Astron. Soc.* 353, 457 (2004).

[5] V. Sahni and A. Starobinsky, *Int. J. Mod. Phy. D* 9, 373 (2000); P. J. Peebles and B. Ratra, *Rev. Mod. Phys.* 75, 559 (2003).

[6] P. J. Steinhardt, *Critical Problems in Physics* (1997), Princeton University Press.

[7] J. Sola and H. Stefancic, *Phys. Lett. B* 624, 147 (2005); J. Sola and H. Stefancic, *Mod. Phys. Lett. A* 21, 479 (2006); I. L. Shapiro and J. Sola, *Phys. Lett. B* 682, 105 (2009).

[8] B. Ratra and P. J. E. Peebles, *Phys. Rev. D* 37, 3406 (1988); C. Wetterich, *Nucl. Phys. B* 302, 668 (1988).

[9] R. R. Caldwell, Phys. Lett. B 545, 23 (2002); R. R. Caldwell, M. Kamionkowski and N. N. Weinberg, *Phys. Rev. Lett.* 91, 071301 (2003).

[10] B. Feng, X. L. Wang and X. M. Zhang, *Phys. Lett. B* 607, 35 (2005).

[11] S. D. H. Hsu, *Phys. Lett. B* 594, 13 (2004); M. Li, *Phys. Lett. B* 603, 1 (2004); Q. G. Huang and M. Li, *JCAP* 0408, 013 (2004);

[12] P. Horava, *Phys. Rev. D* 79, 084008 (2009); G. Calcagni, arXiv:0904.0829 [hep-th]; E. Kiritsis and G. Kofinas, *Nucl. Phys. B* 821, 467 (2009);

[13] C. Wetterich, *Astron. Astrophys.* 301, 321 (1995) L. Amendola, *Phys. Rev. D* 60, 043501 (1999);

[14] Z. K. Guo, R. G. Cai and Y. Z. Zhang, *JCAP* 0505, 002 (2005); Z. K. Guo and Y. Z. Zhang, *Phys. Rev. D* 71, 023501 (2005);

[15] H. Garcia-Compean, G. Garcia-Jimenez, O. Obregon and C. Ramirez, *JCAP* 0807, 016 (2008); M. Jamil and M. A. Rashid, *Eur. Phys. J. C* 60, 141 (2009).

[16] B. Wang, Y. G. Gong and E. Abdalla, *Phys. Lett. B* 624, 141 (2005);

[17] H. Kim, H. W. Lee and Y. S. Myung, *Phys. Lett. B* 632, 605 (2006).

[18] M. R. Setare, *JCAP* 0701, 023 (2007).

[19] J. Grande, J. Sola and H. Stefancic, *JCAP* 0608, 011 (2006); J. Grande, A. Pelinson and J. Sola, *Phys. Rev. D* 79, 043006 (2009).

[20] N. Cruz, S. Lepe and F. Pena, *Phys. Lett. B* 663, 338 (2008).

[21] A. Berera, *Phys. Rev. Lett.* 75, 3218 (1995);

[22] N. Arkani-Hamed, L. J. Hall, C. F. Kolda and H. Murayama, *Phys. Rev. Lett.* 85, 4434 (2000).

[23] L. Amendola, Phys. Rev. D 62 (2000) 043511; L. P. Chimento et al, *Phys. Rev. D* 67 (2003) 08351.

[24] E. W. Kolb, and M. S. Turner, *The Early Universe* (Westview Press, Boulder, 1994).

[25] W. Zimdahl, and D. Pavon, Phys. Lett. B 521, 133 (2001); L. P. Chimento, A. S. Jakubi, D. Pavon, and W. Zimdahl, *Phys. Rev. D* 67, 083513 (2003).

[26] H. M. Sadjadi, and M. Honardoost, *Phys. Lett. B* 647 (2007) 231; H. M. Sadjadi, and M. Alimohammadi, *Phys. Rev. D* 74 (2006) 103007; H. M. Sadjadi, *Phys. Lett. B* 687 (2010) 114; H. M. Sadjadi, *Eur. Phys. J. C* 66 (2010) 445.

[27] P. Horava, JHEP 0903, 020 (2009); P. Horava, *Phys. Rev. Lett.* 102, 161301 (2009).

[28] R. G. Cai, B. Hu and H. B. Zhang, *Phys. Rev. D* 80, 041501. (2009).

[29] G. Calcagni, *JHEP* 0909, 112 (2009).

[30] H. Wei and R. G. Cai, *Phys. Lett. B* 660, 113 (2008).

[31] M. Li, *Phys. Lett. B* 603, 1 (2004).

[32] E. Komatsu et al., arXiv:1001.4538 [astro-ph.CO].

[33] S. Dutta and E. N. Saridakis, arXiv:1002.3373 [hep-th].

[34] H. Wei and R. G. Cai, *Phys. Lett. B* 663, 1 (2008).

[35] Z. K. Guo, N. Ohta and S. Tsujikawa, *Phys. Rev. D* 76, 023508 (2007).

[36] C. Feng, B. Wang, E. Abdalla and R. K. Su, *Phys. Lett. B* 665, 111 (2008).

[37] B. Wang, J. Zang, C. Y. Lin, E. Abdalla and S. Micheletti, *Nucl. Phys. B* 778, 69 (2007).

[38] A. Ashtekar and E.Wilson-Ewing, arXiv:0805.3511 [gr-qc]; M. Bojowald, Living Rev. Rel. 8, 11 (2005); R-G. Cai, L-M Cao, Y-P. Hu, *JHEP* 0808, 090 (2008); S. Li and Y. Ma, *Eur. Phys. J. C* 68, 227 (2010); M. Jamil and U. Debnath, Astrophys. *Space Sci.* 333, 3 (2011).

[39] A. Ashtekar, J. Baez, A. Corichi and K. Krasnov, *Phys. Rev. Lett.* 80, 904 (1998).

[40] Yi-Fu Cai, J. Liu, H. Li, *Phys. Lett. B* 690, 213 (2010); S. Nojiri, S. D. Odintsov , *Phys. Lett. B* 652, 343 (2007).

[41] A. Ashtekar, T. Pawlowski and P. Singh, *Phys. Rev. Lett.* 96, 141301 (2006); A. Ashtekar, T. Pawlowski and P. Singh, *Phys. Rev. D* 74, 084003 (2006); A. Ashtekar, T. Pawlowski, P. Singh, K. Vandersloot, *Phys. Rev. D* 75, 024035 (2007).

[42] S. Chen, J. Jing, *Class. Quant. Grav.* 26, 155006 (2009).

[43] M. Jamil, D. Momeni, R. Myrzakulov, *Eur. Phys. J. C* 71, 1711 (2011)

[44] M. Jamil, D. Momeni, R. Myrzakulov, *Eur. Phys. J. C* 72, 2137 (2012); M. Jamil, D. Momeni, R. Myrzakulov, *Eur. Phys. J. C* 73, 2267 (2013).

[45] R.Ferraro, F.Fiorini, *Phys. Rev. D* 75, 084031 (2007); R. Ferraro, F. Fiorini, *Phys. Rev. D* 78, 124019 (2008).

[46] K. Bamba, R. Myrzakulov, S. Nojiri, S. D. Odintsov, arXiv:1202.4057v2 [physics.gen-ph].

[47] M. Jamil, D. Momeni, R. Myrzakulov, *Eur. Phys. J. C* 72, 1959 (2012).

[48] M. Jamil, E.N. Saridakis, M.R. Setare, *Phys. Rev. D* 81, 023007 (2010).

[49] M. Jamil, D. Momeni, R. Myrzakulov, *Eur. Phys. J. C* 72, 2122 (2012).

[50] M. H. Daouda, M. E. Rodrigues, M. J. S. Houndjo, *Eur. Phys. J. C* 72, 1890 (2012).

[51] A. Boyarsky, A. Neronov, O. Ruchayskiy, I. Tkachev, *Phys. Rev. Lett.* 104, 191301 (2010); N. Fornengo, R. Lineros, M. Regis, M. Taoso, *Phys. Rev. Lett.* 107, 271302 (2011).

In: Research Development in Dark Matter Research ISBN: 978-1-62948-010-7
Editors: Nori Kinjo and Akira Nakajima

Chapter 6

ROTATION CURVES IN BOSE-EINSTEIN CONDENSATE DARK MATTER HALOS

M. Dwornik, Z. Keresztes and L. Á. Gergely*
Department of Theoretical Physics,
University of Szeged, Szeged, Hungary
Department of Experimental Physics,
University of Szeged, Szeged, Hungary

Abstract

The study of the rotation curves of spiral galaxies reveals a nearly constant cored density distribution of Cold Dark Matter. N-body simulations however lead to a cuspy distribution on the galactic scale, with a central peak. A Bose-Einstein condensate (BEC) of light particles naturally solves this problem by predicting a repulsive force, obstructing the formation of the peak. After succinctly presenting the BEC model, we test it against rotation curve data for a set of 3 High Surface Brightness (HSB), 3 Low Surface Brightness (LSB) and 3 dwarf galaxies. The BEC model gives a similar fit to the Navarro-Frenk-White (NFW) dark matter model for all HSB and LSB galaxies in the sample. For dark matter dominated dwarf galaxies the addition of the BEC component improved more upon the purely baryonic fit than the NFW component. Thus despite the sharp cut-off of the halo density, the BEC dark matter candidate is consistent with the rotation curve data of all types of galaxies.

*E-mail address: gergely@physx.u-szeged.hu

PACS: 05.45-a, 52.35.Mw, 96.50.Fm

Keywords: Dark matter halos

1. Introduction

Cosmological observations provide compelling evidence that about 95% of the content of the Universe resides in the unknown dark matter and dark energy components [60]. The former resides in bound systems as non-luminous matter [57, 68], the latter in the form of a zero-point energy pervading the whole Universe. Dark matter is thought to be composed of pressureless, cold, neutral, weakly interacting massive particles, beyond those existing in the Standard Model of Particle Physics, and not yet detected in accelerators or in dedicated direct and indirect searches, excepting gravitational has been found. Therefore the possibility that the Einstein (and Newtonian) theory of gravity breaks down at the scale of galaxies cannot be excluded a priori. While some studies show that the luminous matter alone can explain the rotation in the innermost galactic regions [20,54,70], dark matter is still required on the larger scale. Several theoretical models, based on a modification of Newton's law or of general relativity, have been proposed to explain the behavior of the galactic rotation curves [5–8, 46, 50, 51, 64, 67]. In brane world models, the galactic rotation curves can be naturally explained without introducing dark matter [24, 45, 62]. There is also a possibility that the rotation of galaxies in the outermost regions could be driven by magnetic fields, rather than dark matter [3].

In recent cosmological models the primordial density fluctuations are generated during an inflationary period and they are the seeds of the bottom-up structure formation model. The post-inflation regime is usually described by the ΛCDM (cosmological constant + Cold Dark Matter) model which is consistent with the vast majority of the available observations, including the large scale matter distribution, the Ia type supernovae observations and the temperature fluctuations in the cosmic microwave background radiation [55, 56, 61].

However, the investigation of spiral galaxies clearly shows that the mass distribution of galactic-scale objects can not be explain satisfactory within the framework of the ΛCDM cosmological model. The predicted halo density profile is approximately isothermal over a large range in radii, and it shows a well pronounced central cusp [52]. The Navarro-Frenk-White (NFW) density profile is proportional to $1/r$ close to the centre. On the observational side however, high-resolution rotation curves show instead that the actual distribution of dark matter is much shallower [11], presenting a constant density core. The Burkert density profile shows a correlation between the enclosed surface densities of luminous and dark matter in galaxies [23]. However the astrophysical origin of this empirical density distribution remains unaddressed.

The knowledge of the mass distribution of spiral galaxies is a crucial step in the search for non-baryonic dark matter. Most of the present models assume that, beside the stellar disk and bulge, there is a spherically symmetric, massive dark matter halo, which dominates the total galaxy mass and also determines the dynamics of the stellar disk at the outer regions. Nevertheless Ref. [37] found that the mass distribution for some galaxies cannot be spherical at larger radii and instead a flattened mass distribution (global disk model) better approximates the gravitational potential.

In this chapter we consider scalar field dark matter halos which have undergone a Bose-Einstein condensation (BEC) [6, 13, 14, 21, 33, 34, 63, 72]. Below a critical temperature, bosons favor joining highly populated low-energy states. At the end of this process, bosons will occupy the same quantum ground state and form a coherent matter wave, the Bose-Einstein condensate. In this state, the bosons exhibit a repulsive interaction which prevents the formation of central density cusps by gravitational attraction.

This chapter is organized as follows. The basic properties of the Bose-Einstein condensed dark matter are reviewed in Section 2. Then the theoretical predictions of the model are compared with the observed rotation curve data of several types of galaxies (High Surface Brightness, Low Surface Brightness and Dwarf Galaxies, respectively), in Section 3. We discuss the results in Section 4.

2. The Bose-Einstein Condensate

An ideal Bose gas (a cloud of non-interacting bosons) confined in a box of volume $V = L^3$ obeys the Bose-Einstein thermal distribution $f = \{\exp\left[\left(\epsilon - \mu\right)/k_B T\right] - 1\}^{-1}$, where $\epsilon = p^2/2m$ stands for the energy of the bosons, determined by their mass m and magnitude p of their 3-momentum vector, μ is the chemical potential, T the temperature and k_B the Boltzmann constant. The 3-momentum $\mathbf{p} = 2\pi\hbar\mathbf{q}/L$ is discretized through the dimensionless vector $\mathbf{q}$, with integer components. The lowest energy state has $p = 0$. The number of uncondensated bosons at temperature T is

$$N_T = \sum_{p \neq 0} \frac{1}{\exp\left[\left(p^2/2m - \mu\right)/k_B T\right] - 1} . \tag{1}$$

If the energy levels follow each other densely, i.e. the the thermal energy is much larger than the smallest energy spacing between the single-particle levels:

$$k_B T \gg \frac{2\pi^2 \hbar^2}{m V^{2/3}} , \tag{2}$$

the summation in Eq. (1) can be replaced by an integral over the momentum ($\sum_{\mathbf{p}} \rightarrow V \int d\mathbf{p}/\left(2\pi\hbar\right)^3$) cf. Ref. [59], thus:

$$N_T = \frac{V}{\lambda_T^3} g_{3/2}\left[\exp\left(\mu/k_B T\right)\right] . \tag{3}$$

Here

$$\lambda_T = \hbar \sqrt{\frac{2\pi}{m k_B T}} \tag{4}$$

is the thermal de Broglie wavelength and

$$g_{3/2}\left(z\right) = \frac{2}{\sqrt{\pi}} \int_0^\infty dx \frac{\sqrt{x}}{e^x/z - 1} \tag{5}$$

is a special Bose function, with $z = \exp\left(\mu/k_B T\right)$ and the integral variable $x = p^2/2mk_B T$. At such temperatures when the ground state is vacant, the value of the chemical potential can be expressed from the relation $N_T = N$, with N the total number of bosons. If the lowest energy state $\epsilon_0 = 0$ is occupied, the chemical potential is given by

$$\mu = -k_B T \ln\left(1 + \frac{1}{n_0}\right) \approx -\frac{k_B T}{n_0} , \tag{6}$$

with the ground state particle number n_0. In the thermodynamic limit $n_0, N, V \rightarrow \infty$, $n = N/V$, n_0/V →const, such that n_0/N can be expressed as

$$\frac{n_0}{N} = 1 - \left(\frac{T}{T_c}\right)^{3/2} \quad \left(T \leq T_c\right) . \tag{7}$$

The critical temperature

$$T_c = \frac{2\pi\hbar^2}{mk_B}\left(\frac{n}{g_{3/2}(0)}\right)^{2/3}, \tag{8}$$

with $g_{3/2}(0) = 2.612$, is typically low ($T_c = 3.13$ K for ^{4}He liquid at saturated vapour pressure), and represents the temperature below which the bosons start to condensate into the lowest energy state. By further decreasing the temperature, the relative number of particles in the ground state increases. For $T > T_c$ the ground state is vacant.

The condition for the condensation $T < T_c$ can be rewritten as a relation between the average distance of the bosons $l = \sqrt[3]{V/N} = n^{-1/3}$ and their thermal de Broglie wavelength as

$$l < \frac{\lambda_T}{\zeta^{1/3}} \approx 0.73\lambda_T\,. \tag{9}$$

With the bosons considered a quantum-mechanical wave packet of the order of its de Broglie wavelength, the condensation occurs at the low temperatures where their wavelengths overlap.

As the thermodynamic limit is never realized exactly, corrections arising from the finite size slightly alter the value of the critical temperature, for details see [29], [39], [40], [32].

In a dense, non-ideal (self-interacting) Bose gas the particles can form molecules, and they can reach a more stable state than a BEC. Two atoms can form a molecule if a third particle takes momentum away. In a dilute gas such a scenario can be avoided. Therefore BEC can be formed in a dilute and ultracold Bose gases. The gas is considered dilute if the characteristic length of the interaction l_{int} is much smaller than the average distance of the bosons, thus $l_{int}^3 n \ll 1$. In a dilute gas the bosons are weakly interacting through two-particle interactions. BEC can form in a dilute, non-ideal Bose gas, however the condensate fraction is smaller and the critical temperature is again altered [25], [26] , [69], [15]. Experimentally, BEC has been realized first by different groups in ^{87}Rb ([1], [30], [19]), and in ^{23}Na ([16], [31]), in ^{7}Li ([10]).

2.1. Mean Field Approximation, the Gross-Pitaevskii Equation

The static configuration of N interacting scalar bosons placed in the external potential V_{ext} in a second quantized formalism is characterized [15] by the Hamiltonian operator

$$\begin{aligned}\hat{H} &= \int d\mathbf{r}'\hat{\Psi}^{+}\left(\mathbf{r}'\right)\left[-\frac{\hbar^{2}}{2m}\Delta' + V_{ext}\left(\mathbf{r}'\right)\right]\hat{\Psi}\left(\mathbf{r}'\right) \\ &+\frac{1}{2}\int d\mathbf{r}'d\mathbf{r}''\hat{\Psi}^{+}\left(\mathbf{r}'\right)\hat{\Psi}^{+}\left(\mathbf{r}''\right)V_{self}\left(\mathbf{r}'-\mathbf{r}''\right)\hat{\Psi}\left(\mathbf{r}'\right)\hat{\Psi}\left(\mathbf{r}''\right) . \quad (10)\end{aligned}$$

The operators (singled out by hats) are taken in the *Schrödinger picture.*[1] The boson field operators $\hat{\Psi}\left(\mathbf{r}'\right)$ and $\hat{\Psi}^{+}\left(\mathbf{r}'\right)$ annihilate and create a particle at the position $\mathbf{r}'$, while Δ' is the 3-dimensional Laplacian with respect to the coordinates $\mathbf{r}'$. The *repulsive*, two-body interatomic potential is

$$V_{self} = \lambda\delta\left(\mathbf{r}-\mathbf{r}'\right) \quad (11)$$

with

$$\lambda = \frac{4\pi\hbar^{2}a}{m} \quad (12)$$

a self-coupling constant, given in terms of the scattering length a.

The field operator $\hat{\Psi}\left(\mathbf{r}\right)$ is decomposed in terms of the single-particle annihilation operators $\hat{a}_{\alpha}$ as

$$\hat{\Psi}\left(\mathbf{r}\right) = \sum_{\alpha}\Psi_{\alpha}\left(\mathbf{r}\right)\hat{a}_{\alpha} , \quad (13)$$

where Ψ_{α} is the wave function of single-particle state $|\alpha\rangle$. The summation is taken over the single-particle state. The functions Ψ_{α} are orthonormal and form a complete set of single-particle wave functions, i.e.

$$\sum_{\alpha}\Psi_{\alpha}\left(\mathbf{r}\right)\Psi_{\alpha}^{*}\left(\mathbf{r}'\right) = \delta\left(\mathbf{r}-\mathbf{r}'\right) , \quad (14)$$

where a star denotes the complex conjugation. Denoting the particle numbers in some state (labeled α) by n_{α}, the bosonic annihilation $\hat{a}_{\alpha}$ and creation $\hat{a}_{\alpha}^{+}$ operators act on the Fock space as

$$\begin{aligned}\hat{a}_{\alpha}\left|n_{0},n_{1},...,n_{\alpha},...\right\rangle &= \sqrt{n_{\alpha}}\left|n_{0},n_{1},...,n_{\alpha}-1,...\right\rangle , \quad (15)\\ \hat{a}_{\alpha}^{+}\left|n_{0},n_{1},...,n_{\alpha},...\right\rangle &= \sqrt{n_{\alpha}+1}\left|n_{0},n_{1},...,n_{\alpha}+1,...\right\rangle , \quad (16)\end{aligned}$$

[1]The Hamiltonian operator coincides in the Schrödinger and Heisenberg pictures since it does not depend explicitly on time.

and satisfy the following commutation relations:

$$\left[\hat{a}_\alpha, \hat{a}_\beta^+\right] = \delta_{\alpha\beta}\ , \ \left[\hat{a}_\alpha, \hat{a}_\beta\right] = 0\ , \ \left[\hat{a}_\alpha^+, \hat{a}_\beta^+\right] = 0\ . \tag{17}$$

The numbers n_α are the eigenvalues of the operator $n_\alpha = \hat{a}_\alpha^+ \hat{a}_\alpha$. The commutation relations

$$\left[\hat{\Psi}\left(\mathbf{r}\right), \hat{\Psi}^+\left(\mathbf{r}'\right)\right] = \delta\left(\mathbf{r}-\mathbf{r}'\right)\ , \ \left[\hat{\Psi}\left(\mathbf{r}\right), \hat{\Psi}\left(\mathbf{r}'\right)\right] = 0\ , \ \left[\hat{\Psi}^+\left(\mathbf{r}\right), \hat{\Psi}^+\left(\mathbf{r}'\right)\right] = 0 \tag{18}$$

for the field operators follow from (17) and (14).

Since the Hamiltonian operator is time-independent, the boson field operator $\hat{\Psi}\left(\mathbf{r}, t\right)$ in the *Heisenberg picture* is

$$\begin{aligned} \hat{\Psi}\left(\mathbf{r}, t\right) &= e^{i\hat{H}t/\hbar}\hat{\Psi}\left(\mathbf{r}\right)e^{-i\hat{H}t/\hbar} = \sum_\alpha \Psi_\alpha\left(\mathbf{r}\right)e^{i\hat{H}t/\hbar}\hat{a}_\alpha e^{-i\hat{H}t/\hbar} \\ &= \sum_\alpha \Psi_\alpha\left(\mathbf{r}\right)\hat{a}_\alpha\left(t\right)\ , \end{aligned} \tag{19}$$

with

$$\hat{a}_\alpha\left(t\right) = e^{i\hat{H}t/\hbar}\hat{a}_\alpha e^{-i\hat{H}t/\hbar}\ . \tag{20}$$

Similar equations are valid for the respective adjoint operators $\hat{\Psi}^+\left(\mathbf{r}, t\right)$, $\hat{a}_\alpha^+\left(t\right)$. The field operators obeys the Heisenberg equation, i.e.:

$$\begin{aligned} i\hbar\frac{\partial}{\partial t}\hat{\Psi}\left(\mathbf{r}, t\right) &= \left[\hat{\Psi}\left(\mathbf{r}, t\right), \hat{H}\right] \\ &= \left[-\frac{\hbar^2}{2m}\Delta + V_{ext}\left(\mathbf{r}\right) + \lambda\hat{\Psi}^+\left(\mathbf{r}, t\right)\hat{\Psi}\left(\mathbf{r}, t\right)\right]\hat{\Psi}\left(\mathbf{r}, t\right) \end{aligned} \tag{21}$$

Instead Eq. (21), it is effective computationally to use a *mean-field approximation*. The basic idea [9] is to separate the BEC contribution in the field operator as

$$\hat{\Psi}\left(\mathbf{r}, t\right) = \Psi_0\left(\mathbf{r}\right)\hat{a}_0\left(t\right) + \hat{\Psi}'\left(\mathbf{r}, t\right)\ , \tag{22}$$

where the zero subscript denotes the ground-state and $\hat{\Psi}'\left(\mathbf{r}, t\right)$ carries the effects of the excited states. BEC occurs when the number of particles $n_0\left(t\right)$ in the condensate becomes very large, hence the states with $n_0\left(t\right)$ and $n_0\left(t\right)+1$ correspond to the same configuration. In this case $\hat{a}_0\left(t\right) \approx \hat{a}_0^+\left(t\right) \approx \sqrt{n_0\left(t\right)}$ and the expectation value of the BEC contribution is

$$\psi(\mathbf{r}, t) = \sqrt{n_0\left(t\right)}\Psi_0\left(\mathbf{r}\right)\ , \tag{23}$$

the wave function of the condensate. By comparison the contribution from the non-condensed part is small, therefore $\hat{\Psi}'(\mathbf{r},t)$ represents a perturbation with a negligible expectation value in the leading order approximation.

The probability density

$$\rho(\mathbf{r},t) = |\psi(\mathbf{r},t)|^2 , \tag{24}$$

by the choice $\int d\mathbf{r} \, |\Psi_0(\mathbf{r})|^2 = 1$ is normalized to

$$n_0(t) = \int d\mathbf{r}\rho(\mathbf{r},t) , \tag{25}$$

such that $\rho(\mathbf{r},t)$ also represents the number density of the condensate.

In the leading order approximation (neglecting the contribution of the excited states) Eq. (21) becomes

$$i\hbar\frac{\partial}{\partial t}\psi(\mathbf{r},t) = \left[-\frac{\hbar^2}{2m}\Delta + V_{ext}(\mathbf{r}) + \lambda\rho(\mathbf{r},t)\right]\psi(\mathbf{r},t) , \tag{26}$$

known as the Gross-Pitaevskii equation [27], [28], [58] which describes the Bose-Einstein condensate in the mean-field approximation.

2.2. Madelung Hydrodynamic Equations

In order to find a solution of Eq. (26) it is worth to use the Madelung representation of complex wave-functions [43], [73]:

$$\psi(\mathbf{r},t) = \sqrt{\rho(\mathbf{r},t)}\exp\left[\frac{i}{\hbar}S(\mathbf{r},t)\right] , \tag{27}$$

where the real-valued phase $S(\mathbf{r},t)$ has the dimension of an action. The real part of Eq. (26) gives

$$\frac{\partial S}{\partial t} + \frac{1}{2m}(\nabla S)^2 + \lambda\rho + V_{ext} + V_Q = 0 , \tag{28}$$

corresponding to a generalized Hamilton-Jacobi equations with quantum correction potential

$$V_Q = -\frac{\hbar^2}{2m}\frac{\Delta\sqrt{\rho}}{\sqrt{\rho}} . \tag{29}$$

The imaginary part of (26) becomes a continuity equation

$$\frac{\partial \rho}{\partial t} + \nabla \left(\rho \mathbf{v}\right) = 0\,, \tag{30}$$

while the gradient of (28) gives

$$m\rho \left[\frac{\partial \mathbf{v}}{\partial t} + \left(\mathbf{v}\cdot\nabla\right)\mathbf{v}\right] = -\nabla p - \rho \nabla V_{ext} - \rho \nabla V_Q\,. \tag{31}$$

Here we have introduced the notations

$$\mathbf{v} = \frac{\nabla S\left(\mathbf{r}\right)}{m}\,, \tag{32}$$

and

$$p = \frac{\lambda}{2}\rho^2\,. \tag{33}$$

The ith component of the last term of Eq. (31) can be rewritten as [2]

$$\rho \nabla_i V_Q = \sum_j \nabla_j \sigma_{ij}^Q\,, \tag{34}$$

where

$$\sigma_{ij}^Q = -\frac{\hbar^2}{4m}\rho \nabla_i \nabla_j \ln \rho\,. \tag{35}$$

Eqs. (30) and (31) correspond to the usual continuity and Euler equations of fluid mechanics, with $\mathbf{v}$ the classical velocity field, p the pressure and σ_{ij}^Q representing a quantum correction to the stress tensor. Eqs. (30) and (31) are the Madelung hydrodynamic equations.

2.3. Self-Gravitating, Spherically Symmetric BEC Distribution in the Thomas-Fermi Approximation

A stationary state ψ is

$$\psi(\mathbf{r}, t) = \sqrt{\rho\left(\mathbf{r}\right)} \exp\left(\frac{i\mu}{\hbar}t\right)\,, \tag{36}$$

with μ =const. Then the continuity equation is automatically satisfied while Eq. (28) leads to

$$V_{ext} + V_Q + \lambda\rho = \mu\,. \tag{37}$$

The quantum correction potential V_Q has significant contribution only close to the bound [75], therefore it can be neglected as compared to the self-interaction term $\lambda\rho$. This *Thomas-Fermi approximation* becomes increasingly accurate with an increasing number of particles [42].

If $V_{ext}(\mathbf{r})/m$ is the Newtonian gravitational potential created by the condensate, it satisfies the Poisson equation:

$$\Delta \frac{V_{ext}}{m} = 4\pi G \rho_{BEC} , \tag{38}$$

where $\rho_{BEC} = m\rho$ is the mass density of the BEC and G is the gravitational constant.

The Laplacian of Eq. (37) and Eq. (38) give

$$\Delta\rho_{BEC} + \frac{4\pi G m^2}{\lambda}\rho_{BEC} = 0 . \tag{39}$$

For a spherical symmetric distribution this simplifies to

$$\frac{d^2 (r\rho_{BEC})}{dr^2} + \frac{4\pi G m^2}{\lambda}(r\rho_{BEC}) = 0 , \tag{40}$$

with the solution [75], [6]:

$$\rho_{BEC}(r) = \rho_{BEC}^{(c)} \frac{\sin kr}{kr} , \tag{41}$$

where

$$k = \sqrt{\frac{Gm^3}{\hbar^2 a}} \tag{42}$$

and $\rho_{BEC}^{(c)} \equiv \rho_{BEC}(0)$ is a central density, determined from the normalization condition (21) as

$$\rho_{BEC}^{(c)} = \frac{n_0 m k^3}{4\pi^2} . \tag{43}$$

The exists of a central finite density is exactly the required feature which represents an advantage over cuspy dark matter profiles derived from N-body simulation.

At the end of this subsection we comment on the validity of the Thomas-Fermi approximation. The quantum correction potential (29), rewritten in spherical coordinates and inserted into Eq. (37) generates constant and ρ_{BEC}^{-2} terms, while the contribution of the self-interaction term is proportional to ρ_{BEC}. Close to the boundary R_{BEC} therefore the Thomas-Fermi approximation fails. Multiplying Eq. (37) by $\left(\rho_{BEC}/\rho_{BEC}^{(c)}\right)^2$, with ρ_{BEC} given by Eq. (41), and integrating on the range $r \in [0, R_{BEC}]$ gives the global weight of these terms as $\hbar^2/km$ for the V_Q term and $\hbar^2 a n_0$ for the self-interaction term, respectively. Therefore the Thomas-Fermi approximation holds valid for $n_0 \gg 1/ka$, a condition also obtained by a different method in Ref. [75].

2.4. BEC Dark Matter Halo

In the following, we investigate the possibility that dark matter halos are Bose-Einstein condensates [72].

The size of the BEC galactic dark matter halo is defined by $\rho(R_{BEC}) = 0$, giving $k = \pi/R_{BEC}$, i.e.

$$R_{BEC} = \pi\sqrt{\frac{\hbar^2 a}{Gm^3}} . \tag{44}$$

The mass profile of the BEC halo is then given by

$$\begin{aligned} m_{BEC}(r) &= 4\pi \int_0^r \rho_{BEC}(r) r^2 dr \\ &= \frac{4\pi\rho_{BEC}^{(c)}}{k^2} r \left(\frac{\sin kr}{kr} - \cos kr\right) . \end{aligned} \tag{45}$$

The contribution of the BEC halo to the velocity profile of the particles moving on circular orbit under Newtonian gravitational force is

$$v^2(r) = \frac{4\pi G\rho_{BEC}^{(c)}}{k^2}\left(\frac{\sin kr}{kr} - \cos kr\right) . \tag{46}$$

This has to be added to the respective baryonic contribution.

3. Confronting the Bec Model with Rotation Curve Data

In order to test the validity of the BEC dark matter model, we confront the rotation curve data of a sample of 3 High Surface Brightness (HSB) galaxies,

3 Low Surface Brightness (LSB) galaxies and 3 dwarf galaxies, with both the NFW dark matter and the BEC density profiles.

The commonly used NFW model is based on the numerical simulations of dark-matter halos in the ΛCDM framework [52]. The mass density profile is given by

$$\rho_{NFW}(r) = \frac{\rho_s}{(r/r_s)\left(1 + r/r_s\right)^2} , \tag{47}$$

where there are two fit parameters ρ_s and r_s.

The mass within a sphere with radius $r = yr_s$ is then given by

$$M(r) = 4\pi\rho_s r_s^3 \left[\ln(1 + y) - \frac{y}{1 + y}\right] , \tag{48}$$

where y is a dimensionless radial coordinate.

3.1. High Surface Brightness Galaxies

We follow the method described in Ref. [24]. In a High Surface Brightness galaxy we decompose the baryonic component into a thin stellar disk and a spherically symmetric bulge. We assume that the mass distribution of the bulge component follows the deprojected luminosity distribution, the proportionality factor being the mass-to-light ratio. We estimate the bulge parameters from a Sérsic $r^{1/n}$ bulge model, fitted to the optical I-band galaxy light profiles.

3.1.1. Bulge Contribution

The surface brightness profile of the spheroidal bulge component of each galaxy is described by a generalized Sérsic function [71]

$$I_b(r) = I_{0,b} \exp\left[-\left(\frac{r}{r_0}\right)^{1/n}\right] , \tag{49}$$

where $I_{0,b}$ is the central surface brightness of the bulge, r_0 is its characteristic radius and n is the shape parameter of the magnitude-radius curve.

The respective mass over luminosity is the mass-to-light ratio, for the Sun being $\gamma_\odot = 5133$ kg W^{-1}. In what follows, the mass-to-light ratio of the bulge σ will be given in units of $\gamma_\odot$ (solar units), while the masses in units of the solar mass $M_\odot = 1.98892 \times 10^{30}$ kg. The radial distribution of visible mass is given

Table 1. The distances (D) and the photometric parameters of the 3 HSB galaxy sample. Bulge parameters: the central surface brightness ($I_{0,b}$), the shape parameter (n), the characteristic radius (r_0) and radius of the bulge (r_b). Disk parameters: central surface brightness ($I_{0,d}^{HSB}$) and length scale (h^{HSB}) of the disk

Galaxy	D	$I_{0,b}$	n	r_0	r_b	$I_{0,d}^{HSB}$	h^{HSB}
	Mpc	mJy/arcsec^2		kpc	kpc	mJy/arcsec^2	kpc
ESO215G39	61.29	0.1171	0.6609	0.78	2.58	0.0339	4.11
ESO322G76	64.28	0.2383	0.8344	0.91	4.50	0.0251	5.28
ESO509G80	92.86	0.2090	0.7621	1.10	4.69	0.0176	11.03

by the radial distribution of light obtained from the bulge-disk decomposition. Thus the mass of the bulge within the projected radius r is proportional to the surface brightness encompassed by this radius:

$$M_b(r) = \sigma \frac{\mathcal{N}(D)}{F_\odot} 2\pi \int_0^r I_b(r) r dr,$$

where $F_\odot(D)$ is the apparent flux density of the Sun at a distance D Mpc, $F_\odot(D) = 2.635 \times 10^{6-0.4 f_\odot}$ mJy , with $f_\odot = 4.08 + 5 \lg(D/1 \text{ Mpc}) + 25$ mag, and

$$\mathcal{N}(D) = 4.4684 \times 10^{-35} D^{-2} \text{ m}^{-2} \text{ arcsec}^2. \tag{50}$$

Therefore the contribution of the bulge to the rotational velocity is

$$v_b^2(r) = \frac{G M_b(r)}{r}. \tag{51}$$

3.1.2. Disk Contribution

In a spiral galaxy, the radial surface brightness profile of the disk exponentially decreases with the radius [22]

$$I_d(r) = I_{0,d}^{HSB} \exp\left(-\frac{r}{h^{HSB}}\right), \tag{52}$$

where $I_{0,d}^{HSB}$ is the disk central surface brightness and h^{HSB} is a characteristic disk length scale. The contribution of the disk to the circular velocity is [22]

$$v_d^2(x) = \frac{GM_D^{HSB}}{2h^{HSB}} x^2 (I_0 K_0 - I_1 K_1), \tag{53}$$

where I_n and K_n are the modified Bessel functions calculated at $x/2 = r/2h^{HSB}$ and M_D^{HSB} is the total mass of the disk.

3.1.3. Confrontation with Rotation Curve Data

Therefore the rotational velocity in a HSB galaxy receives the following contributions

$$v_{tg}^2(x) = v_b^2(x) + v_d^2(x) + v_{DM}^2(x) \, .$$

We confront the BEC+baryonic model with (HI and H_α) rotation curve data of 3 galaxies already employed in Ref. [24], which were extracted from a larger sample given in Ref. [54] by requiring i) sufficient and accurate data for each galaxy and ii) manifest spherical structure of the bulge (no visible rings and bars). For comparison, the NFW+baryonic model is also tested on the same sample, by plotting the respective rotation curves in both models. The results are represented on Fig. 1.

For the HSB galaxy sample we have derived the best fitting values of the baryonic model parameters $I_{0,b}$, n, r_0, r_b, $I_{0,d}^{HSB}$ and h^{HSB} from the available photometric data.

The BEC and NFW parameters were calculated (together with the corresponding baryonic parameters) by fitting these models to the rotation curve data. These are collected in the tables 1 and 2.

3.2. Low Surface Brightness Galaxies

Low Surface Brightness (LSB) galaxies are characterized by a central surface brightness at least one magnitude fainter than the night sky. These galaxies form the most unevolved class of galaxies [36] and have low star formation rates as compared to their HSB counterparts [49]. LSB galaxies show a wide spread of colours ranging from red to blue [53] and represent a large variety of properties and morphologies. Although the most commonly observed LSB galaxies are dwarfs, a significant fraction of LSB galaxies are large spirals [4].

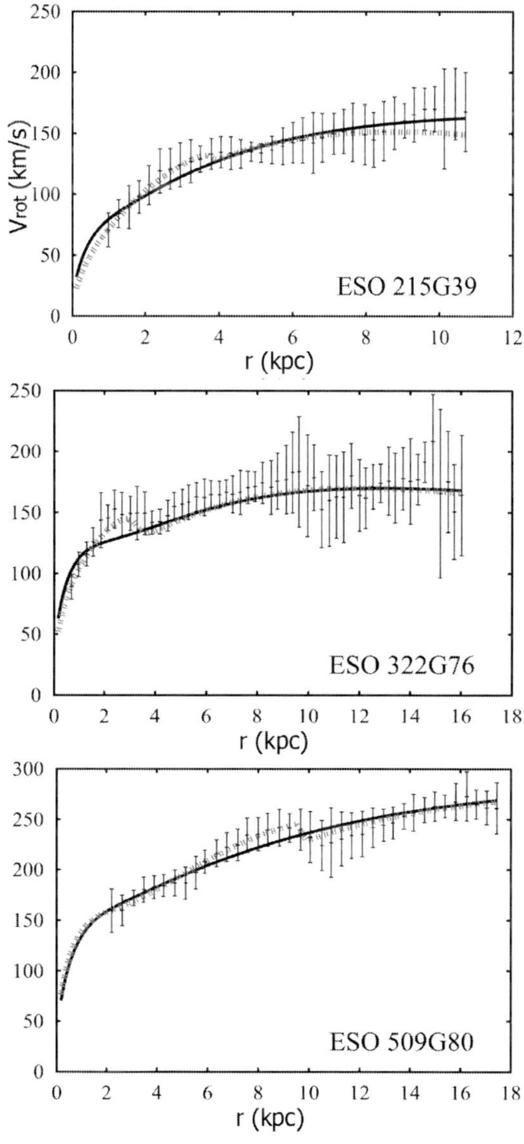

Figure 1. Best fit curves for the HSB galaxy sample. The dashed lines hold in the baryonic matter + BEC model, while the solid lines in the baryonic matter + NFW model. In the second and third example the curves run very close to each other.

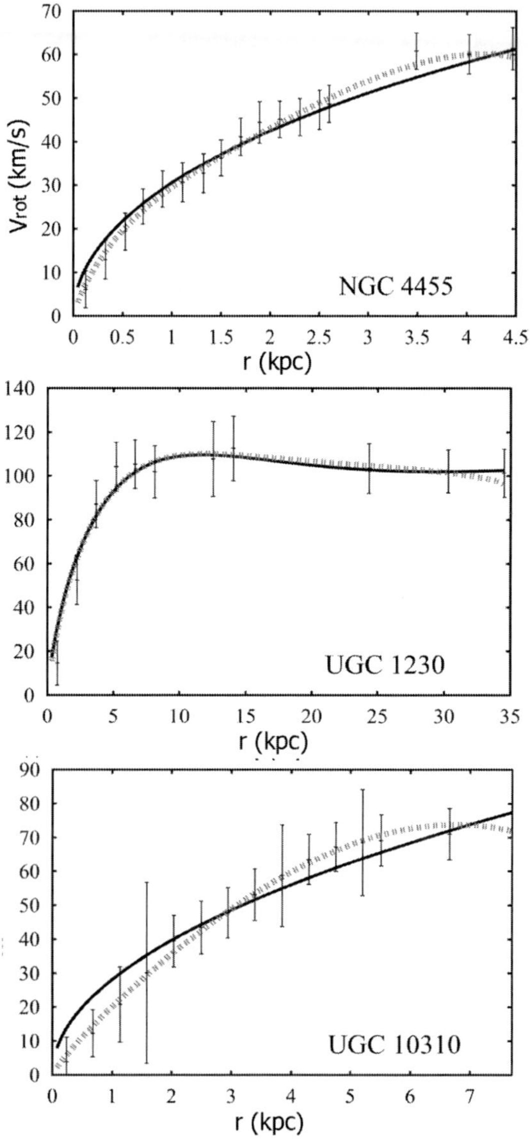

Figure 2. Best fit curves for the 3 LSB galaxies (the dashed lines show the combined BEC+baryonic model, the solid lines show the combined NFW+baryonic profiles).

Table 2. The best fit parameters and the minimum values ($\chi^2_{\min}$) of the χ^2 statistics for the 3 HSB galaxies. Columns 2-5 give the BEC model parameters (radius R_{BEC} and central density $\rho^{(c)}_{BEC}$ of the BEC halo) and the corresponding baryonic parameters (mass-to-light ratio $\sigma\ (BEC)$ of the bulge and total mass of the disk $M^{HSB}_D\ (BEC)$). Columns 7-10 give the NFW model parameters (scale radius r_s and characteristic density ρ_s of the halo) and the corresponding baryonic parameters (mass-to-light ratio $\sigma\ (NFW)$ of the bulge and total mass of the disk $M^{HSB}_D\ (NFW)$). The 1σ confidence levels are shown in the last column (these are the same for both models). For all galaxies $\chi^2_{\min}$ are within the 1σ confidence level. Both model fittings give similar $\chi^2_{\min}$ values

Galaxy	σ(BEC)	M_D^{HSB}(BEC)	R_{BEC}	$\rho^{(c)}_{BEC}$	$\chi^2_{\min}$ (BEC)	σ(NFW)	M_D^{HSB}(NFW)	r_s	ρ_s	$\chi^2_{\min}$(NFW)	1σ
	$\odot$	$10^{10}M_\odot$	kpc	$10^{-21}kg/m^3$		$\odot$	$10^{10}M_\odot$	kpc	$10^{-24}kg/m^3$		
ESO215G39	0.7	4.96	50	0.2	22.9	0.6	3.84	187	14.7	22.22	34.18
ESO322G76	0.8	9.08	5	0.5	48.31	0.9	8.29	920	0.7	49.02	53.15
ESO509G80	1.3	52.02	7	1.6	19.77	0.9	11	22	800	33.48	36.3

Table 3. The best fit BEC and NFW parameters of the 3 LSB galaxies

Galaxy	D	h^{LSB}	$M_D^{LSB}(BEC)$	$\rho^{(c)}_{BEC}$	R_{BEC}	$\chi^2_{\min}(BEC)$	$M_D^{LSB}(NFW)$	ρ_s	r_s	$\chi^2_{\min}(NFW)$	1σ
	Mpc	kpc	$10^9M_\odot$	$10^{-21}kg/m^3$	kpc		$10^9M_\odot$	$10^{-24}kg/m^3$	kpc		
NGC 4455	6.8	0.7	0.236	1.4	5.6	9.4	0	62	39	30.37	18.11
UGC 1230	51	4.5	27	0.1	27.7	19.86	26.2	2	291	19.84	8.17
UGC 10310	15.6	1.9	0.412	1	7.7	2.68	0	3	660	28.78	13.74

Table 4. The best fit BEC+baryonic and NFW+baryonic parameters for the dwarf galaxy sample

Galaxy	$h^{dwarf}(BEC)$	$M_D^{dwarf}(BEC)$	$\rho^{(c)}_{BEC}$	R_{BEC}	$\chi^2_{\min}(BEC)$	$h^{dwarf}(NFW)$	$M_D^{dwarf}(NFW)$	ρ_s	r_s	$\chi^2_{\min}(NFW)$	1σ
	kpc	$10^9M_\odot$	$10^{-21}kg/m^3$	kpc		kpc	$10^9M_\odot$	$10^{-24}kg/m^3$	kpc		
IC 2574	1.2	0.1122	0.4	13	68.47	7.9	28.44	0	0	714.73	44.74
HoI	0.2	0.0107	3.6	1.9	95.26	0.9	0.533	0	0	241.30	20.27
M81dwB nor	0.9	1.023	3.7	0.7	6.19	0.7	0.705	0	0	8.4	10.42

Our model LSB galaxy consists of a thin stellar+gas disk and a cold dark matter component in a form of BEC. The disk component is the same as at the HSB galaxies, the surface brightness profile is [22]

$$I_d(r) = I_{0,d}^{LSB} \exp\left(-\frac{r}{h^{LSB}}\right) ,$$

where $I_{0,d}^{LSB}$ is the central surface brightness and h^{LSB} is the disk length scale. We can calculate the disk contribution to the circular velocity as

$$v_d^2(r) = \frac{GM_D^{LSB}}{2h^{LSB}} x^2 (I_0 K_0 - I_1 K_1), \tag{54}$$

similarly to the case of HSB galaxies.

Therefore for a generic projected radius r, the rotational velocity in this combined model can be written as

$$v_{tg}^2(r) = v_d^2(r) + v_{DM}^2 .$$

We follow the analysis of the BEC model with the rotation curves of 3 LSB galaxies taken from a larger sample [17]. These high quality rotation curve data are based on both HI and $H\alpha$ measurements. From a χ^2-test we have determined the model parameters in both the BEC+baryonic and NFW+baryonic models, these are shown in Table 3. The fitted curves are represented on Fig. 2.

3.3. Dwarf Galaxies

Dwarf galaxies are the most common galaxies in the observable Universe. About 85% of the known galaxies in the Local Volume are dwarfs [38]. Dwarf galaxies are defined as galaxies having an absolute magnitude fainter than $M_B \sim -16$ mag, and more extended than globular clusters [74].

The formation history of dwarf galaxies is not well-understood. According to [76], dwarf galaxies formed at the centers of subhalos orbiting within the halos of giant galaxies. Five main classes of dwarf galaxies are distinguished based on their optical appearance: dwarf ellipticals, dwarf irregulars, dwarf spheroidals, blue compact dwarfs, and dwarf spirals. The last one type can be regarded as the very small end of spirals [48].

All dwarf galaxies have central velocity dispersions $\sim 6 \div 25$ km/s [47]. If the systems are in dynamic equilibrium, the mass derived from these velocity dispersions is much larger than the derived stellar mass. Therefore the dwarf galaxies are among the darkest objects ever observed in the Universe, hence they play an important role in the study of dark matter distribution on small scales. As dwarf galaxies are supposed to be dark matter dominated at all radii, they are ideal objects to prove or falsify various alternative gravity theories [12].

To test the BEC model, we have selected a sample of 3 dwarf galaxies for which high resolution rotation curves are available. We performed the rotation curve fitting with the BEC+baryonic and the NFW+baryonic models, respectively. The baryonic components were the same as in the case of LSB galaxies. However, the length scales of the stellar disks are not available for this sample, therefore they are calculated from χ^2 minimization, too. This comparison allows us to test the viability of our model.

For the investigated dwarf galaxies the best fit BEC and NFW parameters are shown in Table 4 and the fitted rotation curves are represented in Fig. 3.

4. Conclusion

We presented here a comprehensive introduction into the theory of Bose-Einstein condensates (BEC), with emphasis on a spherically symmetric self-gravitating BEC in the Thomas-Fermi approximation.

In the BEC model, large-scale structures like clusters or superclusters of galaxies form similarly as in the Cold Dark Matter model with cosmological constant, thus all predictions of the standard model at large scales are well reproduced [44]. Moreover the BEC model can explain the collisions of galaxy clusters [41] and the acoustic peaks of the cosmic microwave background [66].

Such a BEC has been proposed recently as a galactic dark matter model and has been partially tested by confronting with rotation curve data in Refs. [18,33,65].

Here we have performed a much thorough test of this BEC dark matter model by confronting with a sample of 3 HSB, 3 LSB and 3 dwarf galaxies and also comparing the model predictions with those of the widely accepted NFW dark matter model. We incorporated in all cases realistic baryonic models, taking into account the particularities of the respective galaxy types. Beside the rotation curve data for the HSB galaxies, the surface photometry data was also available. The galaxies have different luminosities, disk length-scales and surface brightnesses. Most of the rotation curve data were densely distributed and uniform in quality.

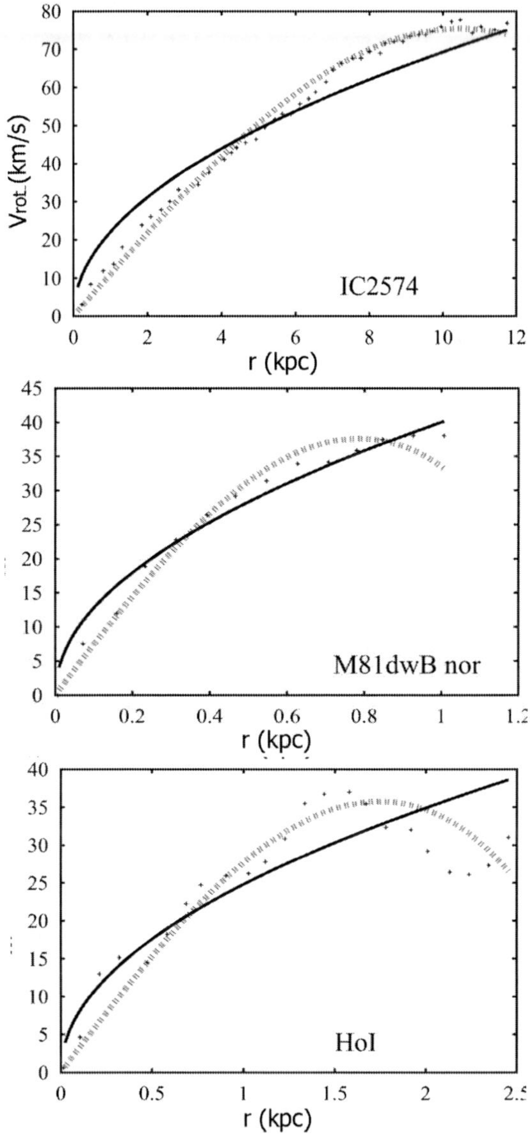

Figure 3. The best fit curves for the dwarf galaxy sample. The BEC+baryonic model (dashed curves) gives a better fit in all cases then the NFW+baryonic model (solid lines). In both cases the fit was performed with the same baryonic model.

We decomposed the circular velocity into its baryonic and dark matter contributions. Then the rotation curves are χ^2 best-fitted with the baryonic parameters and the parameters of the two dark matter halo models (BEC and NFW). For the sample of *HSB galaxies* we found a remarkably good agreement of both dark matter models with the observations. The quality of the fits of the BEC and NFW models with the rotation curve data was comparable. For the sample of *LSB galaxies*, the BEC model gave a slightly better fit than the NFW model. We additionally found that adding the baryonic component results in a better fit than the one presented in [65] for the pure BEC model. For the sample of dark matter dominated *dwarf galaxies*, providing the strongest test of the compared models, the BEC dark matter model gave better results. For all galaxy types we have determined the BEC parameters ρ_{BEC}, R_{BEC}, given in Tables 2, 3 and 4.

We note that due to the sharp cut-off of the BEC halos the very distant behaviour of the universal rotation curves (URCs) [57] is not expected to be reproduced, hence a modification of the BEC model on large distances would be desirable. One suggestion takes into account the effects of the finite dark matter temperature on the properties of the dark matter halos. However it turned out, that these effects do not play a significant role in the description of the dark matter halo density profiles [35]. A possible solution is including vortex lattices in the halo [59, 78]. When a BEC is rotated at a rate exceeding some critical frequency, quantized vortices can be formed. This vortex lattice can influence the galactic rotation curve and provide a flat velocity profile with oscillatory structure [65, 77].

We conclude that the BEC dark matter model is well supported by rotation curve data and it certainly deserves further attention.

References

[1] Anderson, M.H., Ensher, J.R., Matthews, M.R., Wieman, C.E., Cornell, E.A., 1995, *Science* 269, 198.

[2] Barcelo, C., Liberati, S., Visser, M., 2011, *Class. Quant. Grav.* 18, 1137.

[3] Battaner, E., Garrido, J.L., Membrado, M., Florido, E., 1992 *Nature*, 360, 6405, p. 652.

[4] Beijersbergen, M., de Blok, W.J.G., van der Hulst, J.M., 1999, *A&A* 351, 903.

[5] Bertolami, O., Boehmer C.G., Harko, T., Lobo, F.S.N., 2007, *Phys. Rev. D* 75, 104016.

[6] Boehmer, C.G., Harko, T., 2007b, *JCAP* 06, 025.

[7] Boehmer, C.G., Harko, T., Lobo, F.S.N., 2008, *Astropart. Phys.*, 29, 386.

[8] Boehmer, C.G., Harko, T., 2007a, *MNRAS* 379, 2007.

[9] Bogoliubov, N., 1947, *J. Phys. USSR* 11, 23.

[10] Bradley, C.C., Sackett, C.A., Tollett, J.J., Hulet, R.G., 1995, *Phys. Rev. Lett.* 75, 1687.

[11] Burkert, A., 1997, *"Aspects of Dark Matter in Astro-and Particle Physics"*.

[12] Capozziello, S., Cardone, V.F., Troisi, A., 2007, *MNRAS*, 375, 1423.

[13] Chavanis, P.-H., 2011, *Phys. Rev. D* 84, 043531.

[14] Chavanis, P.-H., Delphini, L., 2011, *Phys. Rev. D* 84, 043532.

[15] Dalfovo, F., Giorgini, S., Pitaevskii, L.P., Stringari, S., 1999, *Rev. Mod. Phys.* 71, 463.

[16] Davis, K.B., Mewes, M.-O., Andrews, M.R., van Druten, N.J., Durfee, D.S., Kurn, D.M., Ketterle, W., 1995, *Phys. Rev. Lett.* 75, 3969.

[17] de Blok W.J.G., Bosma A., 2002, *A&A* 385, 816.

[18] Dwornik, M., Keresztes, Z., Gergely, L.Á., 2013, [arXiv:1301.6614].

[19] Ernst, U., Marte, A., Schreck, F., Schuster, J., Rempe, G., 1998, *Europhys. Lett.* 41, 1.

[20] Evans, N.W., 2001, Proc. 3rd InternationalWorkshop on the Identification of DarkMatter.*World Sci.*, Singapore, p. 85.

[21] Fukuyama, T., Morikawa, M., 2006, *Progress of Theoretical Physics* 115, 1047.

[22] Freeman, K.C., 1970, *ApJ* 160, 811.

[23] Gentile, G., Famaey, B., Zhao, H., Salucci, P., 2009, *Nature* 461, 627.

[24] Gergely, L.Á., Harko, T., Dwornik, M., Kupi, G., Keresztes, Z., 2011, *MNRAS* 415, 3275.

[25] Giorgini, S., Pitaevskii, L., Stringari, S., 1996, *Phys. Rev. A* 54, R4633.

[26] Glaum, K., Pelster, A., Kleinert, H., Pfau, T., 2007, *Phys. Rev. Lett.* 98, 080407.

[27] Gross, E.P., 1961, *Nuovo Cimento* 20, 454.

[28] Gross, E.P., 1963, *J. Math. Phys.* 4, 195.

[29] Grossmann, S., Holthaus, M., 1995, Zeit. f. Naturforsch. 50a, 921; *Phys. Lett. A* 208, 188.

[30] Han, D.-J., Wynar, R.H., Courteille Ph., Heinzen D.J., 1998, *Phys. Rev. A* 57, R4114.

[31] Hau, L.V., Busch, B.D., Liu, C. Z. Dutton, Burns, M.M., Golovchenko, J.A., 1998, *Phys. Rev. A* 58, R54.

[32] Haugerud, H., Haugset, T., Ravndal, F., 1997, *Phys. Lett. A* 225, 18.

[33] Harko, T., 2011a, *JCAP* 05, 022.

[34] Harko, T., 2011b, *MNRAS* 413, 3095.

[35] Harko, T., Madarassy, E. J. M., 2011 [arXiv:1110.2829v1].

[36] Impey, C., Bothun, G., 1997, *ARA&A*. 35, 267.

[37] Jalocha, J., Bratek, L., Kutschera, M., Skindzier, P., 2010, *MNRAS* 406, 2805-2816.

[38] Karachentsev, I. D., Karachentseva, V. E., Huchtmeier, W. K., Makarov, D. I., 2004, *AJ* 127, 2031.

[39] Ketterle, W., van Druten N.J., 1996, *Phys. Rev. A* 54, 656.

[40] Kristen, K., Toms, D.J., 1996, *Phys. Rev A* 54, 4188.

[41] Lee, J.W., Lim, S., Choi, D., 2008, [arXiv:0805.3827v1].

[42] Lieb, E.H., Seiringer, R., Yngvason, Y., 2000, *Phys. Rev. A* 61, 043602.

[43] Madelung, E., 1926, *Zeitschrift für Physik* 38, 322.

[44] Magana, J., Matos, T., Robles, V. H., Suarez, A., 2012, *Proceedings of the XIII Mexican Workshop on Particles and Fields* [arXiv:1201.6107v1].

[45] Mak, M. K., Harko, T., 2004, *Phys. Rev. D* 70 , 024010.

[46] Mannheim P. D., 1997, *AJ* 479, 659.

[47] Mateo, M., 1998, *ARA&A* 36, 435.

[48] Matthews, L. D. and Gallagher, J. S., 1997, *AJ* 114, 5.

[49] McGaugh, S.S., 1994, *ApJ* 426, 135.

[50] Milgrom, M., 1983, *APJ* 270, 365.

[51] Moffat, J.W., Sokolov, I.Y., 1996, *Phys. Lett. B* 378, 59.

[52] Navarro, J.F., Frenk, C.S., White, S.D.M., 1996, *ApJ* 462, 563.

[53] O'Neil, K., Bothun, G.D., Schombert, J., Cornell, M.E., Impey, C.D., 1997, *AJ* 144, 244.

[54] Palunas, P., Williams, T.B., 2000, *AJ* 120, 2884.

[55] Padmanabhan, T., 2003, *Phys. Repts.* 380, 235.

[56] Peebles, P. J. E., Ratra, B., 2003, *Rev. Mod. Phys.* 75, 559.

[57] Persic, M., Salucci, P., Stel, F., 1996, *MNRAS* 281, 27.

[58] Pitaevskii, L.P., 1961, *Zh. Eksp. Teor. Fiz.* 40, 646 [Sov. Phys. JETP 13, 451 (1961)].

[59] Pitaevskii, L. P., Stringari S., 2003, *Bose-Einstein Condensation*, Oxford University Press Inc., New York.

[60] Planck collaboration, 2013, [arXiv:1303.5076].

[61] Planck collaboration, 2013, [arXiv:1303.5075].

[62] Rahaman, F., Kalam, M., DeBenedictis, A., Usmani, A.A., Saibal, R., 2008, *MNRAS* 389, 27.

[63] Rindler-Daller, T., Shapiro, P., 2011, [arXiv:1106.1256].

[64] Roberts, M.D., 2004, *Gen. Rel. Grav.* 36, 2423.

[65] Robles, V. H., Matos T., 2012, *MNRAS* 422, 282-289.

[66] Rodriguez-Montoya, I., Magana, J., Matos, T., Perez-Lorenzana, A., 2010, *ApJ* 721, 1509.

[67] Sanders, R.H., 1984, *A&A* 136, L21.

[68] Salucci, P., Persic, M., 1999, *MNRAS* 309, 923.

[69] Schütte, M., Pelster, A., 2008, Critical Temperature of a Bose-Einstein Condensate with 1/r Interactions, *Proceedings of the 9th International Conference*, 23-28, September, 2007, Dresden, Germany, Eds. Janke W. and Pelster A., World Scientific Publishing Co. Pte. Ltd., 2008. ISBN #9789812837271, pp. 417-420.

[70] Sellwood, J.A., Kosowsky, A., 2000, ASP Conf. Ser. Vol. 240, Gas and Galaxy Evolution. *Astron. Soc. Pac.*, San Francisco, p. 311.

[71] Sérsic, J.L., 1968, *Atlas de Galaxias Australes*, Cordoba, Argentina, Observatorio Astronomico.

[72] Sin, S.J., 1994, *Phys. Rev. D* 50, 3650.

[73] Sonego, S., 1991, *Found. Phys.* 21, 1135.

[74] Tammann, G. A., 1994, Dwarf Galaxies in the Past, in Dwarf Galaxies, *ESO Conference and Workshop Proc* No. 49, Eds. G. Meylan and P. Prugniel, Paris, p. 3.

[75] Wang, X.Z., 2001, *Phys. Rev D* 64, 124009.

[76] Wu, X., 2007, submitted to ApJ.

[77] Yu, R. P., Morgan M. J., 2002, *Class. Quantum Grav.* 19, L157.

[78] Zinner, N. T., 2011, [arXiv:1108.4290v1].

In: Research Development in Dark Matter Research ISBN: 978-1-62948-010-7
Editors: Nori Kinjo and Akira Nakajima

Chapter 7

WEYL DARK FLUID FROM THE BRANE WORLD SCENARIO

***M. C. B. Abdalla*[1,*], *P. F. Carlesso*[1,†] *and J. M. Hoff da Silva*[2,‡]**
[1]Instituto de Física Teórica, Universidade Estadual Paulista, Brazil
[2]Faculdade de Engenharia de Guaratinguetá,
Universidade Estadual Paulista, Brazil

Abstract

We review some developments of the Weyl tensor tool in the explanation of the galaxy flat rotation curves within the braneworld context. Besides we also present a qualitative discussion about the sign of the so-called pressure parameter.

Keywords: Braneworld, Weyl Fluid, Galaxy Rotation Curves

1. Introduction

There are several attempts to explain data concerning the apparent necessity of hidden (dark, in this sense) matter in the universe. It would be necessary an extensive review to pinpoint all the indicative data. The first insight towards the dark matter necessity appeared in [1], in the observation of COMA galaxy

*E-mail address: mabdalla@ift.unesp.br
†E-mail address: pablofisico@ift.unesp.br
‡E-mail addresses: hoff@feg.unesp.br; hoff@ift.unesp.br

cluster. Of particular interest for this short review, the analysis of some galaxy rotation curves also indicates a deficit in the necessary mass outside the galaxy core to explain its flat rotation behavior [2, 3, 4].

Generally speaking, the theoretical attempts to explain those phenomena may be separated into two approaches: postulating the existence of an exotic matter field which does not interact with the standard model fields and, on the other hand, gravity modified theories whose geometrical effects may be in order to explain the aforementioned effects. The use of braneworld models in the cosmological context follows this last point of view. Here we shall review some works in the braneworld context used to explain the galaxy flat rotation curves and some generalizations.

Starting from the usual general relativity in 5 dimensions, it is possible to project out the equations to be respected by the gravitational field on the brane [5]. In this procedure, the contribution of the 5-dimensional graviton to the brane gravity is geometrically performed by the appearance of some specific projection of the Weyl tensor. Besides, extra brane stress tensor terms contribute to a new gravitational equation. In this vein, it is the Weyl tensor which may be used for the galaxy flat rotation study. In other words, the contributions of the 5-dimensional graviton could be quite enough to explain that behavior [6, 7, 8, 9].

The work is organized as follows: the next Section gives a short review of the brane induced gravitational effects, expliciting the gravitational equation on the brane, as well as the usual Weyl tensor decomposition. In the Section III we apply the resulting gravitational equation to the spiral galaxies rotation curves, showing how the braneworld gravitational dynamics may solve this problem without any additional dark matter field. Section IV accounts for some qualitative aspects of the brane projected gravitational equation based upon dynamical systems tools. In the last Section we conclude.

2. Brane Gravitational Dynamics

The braneworld scenario used here, inspired by Randall-Sundrum models [10], consists of a 3-brane embedded into a 5-dimensional bulk, with a $\mathbb{Z}_2$ symmetry on the extra spatial dimension. As mentioned in the Introduction, the gravitation in the bulk is performed by the Einstein's field equations. Starting from this hypothesis, we can achieve the brane gravitational equations through the Gauss-Codazzi formalism and the use of some junction conditions. It is well know that the Gauss-Codazzi equations relate the Riemann bulk curvature with the Riemann brane curvature and the junction conditions relate the brane extrinsic curvature with the brane stress-energy tensor $\tau_{\mu\nu}$ [5]. The resulting brane gravitational equations are

$$G_{\mu\nu} = -\Lambda g_{\mu\nu} + \kappa_4^2 \tau_{\mu\nu} + \kappa_5^4 S_{\mu\nu} - E_{\mu\nu}, \tag{1}$$

where $G_{\mu\nu}$ is the Einstein brane tensor, $g_{\mu\nu}$ is the brane metric, $\Lambda = \frac{\kappa_5^2}{2}(\Lambda_5 + \kappa_5^2 \lambda_b^2/6)$ is an effective brane cosmological constant (depending on the bulk cosmological constant Λ_5 and the intrinsic brane tension λ_b), $S_{\mu\nu}$ is a term depending on the brane stress-energy tensor and $E_{\mu\nu}$ is a non-local term, depending on the bulk's Weyl tensor. Moreover, the constants of the eq. (1) are related by the following relations: $\kappa_5^2 = 8\pi G_5$, were G_5 is the bulk gravitational constant, and $\kappa_4^2 = \kappa_5^4 \lambda_b/6$.

The above equations differ from the usual 4-dimensional Einstein's equations by the two last terms on the right-hand side. Such modifications enable us to approach the flat rotation curves problem from the braneworld model point of view. To achieve it, we consider vacuum spherically symmetric solutions of equation (1). The vacuum condition, implies that the $S_{\mu\nu}$ term vanishes. Moreover, for the purpose of this work, we disregard the contribution of the cosmological constant term. The brane gravitational equations become

$$R_{\mu\nu} = -E_{\mu\nu}. \tag{2}$$

Thus, the brane vacuum is related to the 5-dimensional Weyl-projecetd term, that is usually parameterized by a "dark cosmological fluid", the so-called "Weyl fluid" [11]

$$E_{\mu\nu} = -\left(\frac{\kappa_5}{\kappa_4}\right)^4 \left[U\left(u_\mu u_\nu + \frac{h_{\mu\nu}}{3}\right) + 2Q_{(\mu}u_{\nu)} + P_{\mu\nu}\right], \tag{3}$$

where we decompose the metric tensor in terms of a 4-velocity field ($g_{\mu\nu} = h_{\mu\nu} - u_\mu u_\nu$). The above equation is closely related to a cosmological fluid equation where U represents the dark radiation component, $P_{\mu\nu}$ is the anisotropic pressure and Q_μ the energy flux, all of them associate to the dark Weyl fluid.

In order to satisfy the spherical symmetry and the static vacuum condition, we write $U = U(r)$ and $P_{\mu\nu} = P(r)(r_\mu r_\nu - h_{\mu\nu}/3)$, where r_μ is a unit radial vector and r the radial distance. The static vacuum condition also implies that the dark fluid has a vanishing energy flux ($Q_\mu = 0$).

Thought those considerations, we can solve equation (2) for a static and spherically symmetric spacetime, given by the metric

$$ds^2 = -e^{\nu(r)}dt^2 + e^{\lambda(r)}dr^2 + r^2(d\theta^2 + sin^2\theta d\phi^2), \tag{4}$$

where $\nu(r)$ and $\lambda(r)$ are functions of the radial coordinate that will be related with the dark fluid parameters, as we can see by the resolution of (2)

$$-e^{-\lambda}\left(\frac{1}{r^2} - \frac{\lambda'}{r}\right) + \frac{1}{r^2} = 3\alpha U, \tag{5}$$

$$e^{-\lambda}\left(\frac{\nu'}{r} + \frac{1}{r^2}\right) - \frac{1}{r^2} = \alpha(U + 2P), \tag{6}$$

$$\frac{e^{-\lambda}}{2}\left(\nu'' + \frac{\nu'^2}{2} + \frac{\nu' - \lambda'}{r} + \frac{\nu'\lambda'}{2}\right) = \alpha(U - P), \tag{7}$$

$$\nu' = -\frac{U' + 2P'}{2U + P} - \frac{6P}{r(2U + P)}, \tag{8}$$

where the last equation was obtained by taking the covariant derivative of (2). We have yet $\alpha = \frac{16\pi G\kappa_4^4}{\kappa_5^4\lambda_b}$ and the prime denotes derivation with respect to r.

Those solution are similar to the Schwarzschild one, but with the presence of the Weyl fluid parameters. From the equation (5) we can obtain one of the metric components

$$e^{-\lambda} = 1 - \frac{C}{r} - \frac{Q}{r}, \tag{9}$$

where $Q = 3\alpha\int Ur^2dr$, or

$$\frac{dQ}{dr} = 3\alpha Ur^2. \tag{10}$$

Substituting Eq. (8) in (6) and using (9), we obtain the relation

$$\frac{dU}{dr} = -2\frac{dP}{dr} - 6\frac{P}{r} - \frac{(2U + P)\Big(C + Q + \alpha r^3(U + 2P)\Big)}{r^2\Big(1 - \frac{C}{r} - \frac{Q}{r}\Big)}, \tag{11}$$

that will be useful in the next sections.

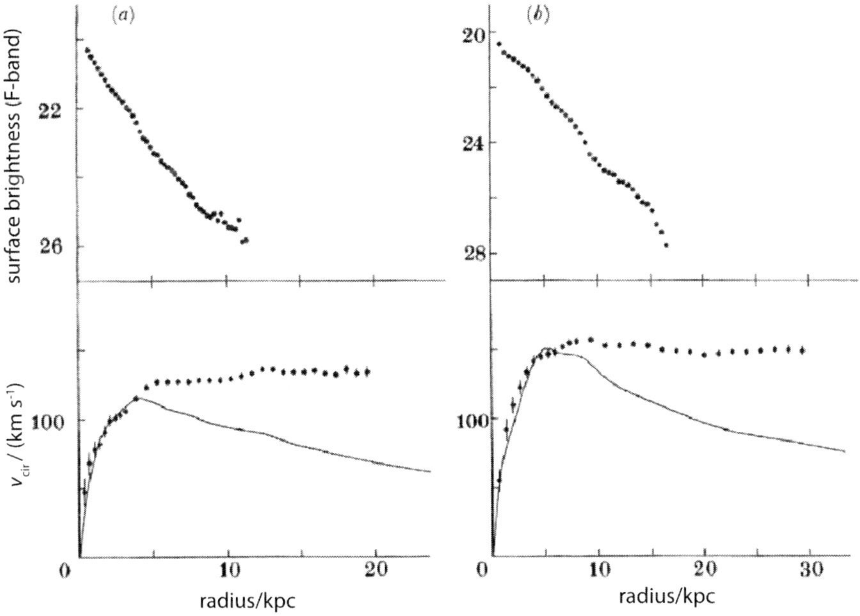

Figure 1. Rotation curves for the galaxies NGC2403 and NGC3198 [3].

3. Spiral Galaxies Rotation Curves

The observed tangential velocity of astrophysical objects (as hydrogen clouds) orbiting spiral galaxies shows a behavior incompatible with the predictions of both general relativity and newtonian gravity[2].

In the above figure, the upper graphics give the luminous matter distribution for each galaxy, the lower graphics give the respective rotation curves, where the solid line represents the theoretical prediction based upon the baryonic matter distribution. The dotted line indicates the experimental data.

The aforementioned discrepancy is usually bypassed with the assumption of the existence of some dark matter halo, whose distribution overcomes the difference between predictions and measurements. There are several dark matter candidates. Nevertheless, at the present date, there is no evidence, from the particle physics data, confirming such hypothesis. An alternative to dark matter is to consider braneworld scenarios, where the modified gravitational dynamics on the brane may be used in the explanation of flat rotation curves. We consider here, an exterior spacetime of a spiral galaxy. Let us assume that the mass distribution is concentred mostly in the central galaxy core, with spherical symmetry. It implies in a vacuum spherical symmetric solution for the brane spacetime outside the core, as discussed in the previous section.

Given the metric (4) for the brane spacetime, it is possible to achieve an expression for the tangential velocity of a test particle in a circular motion about the galaxy core [12]

$$v_{tg}^2 = \frac{r\nu}{2}, \tag{12a}$$

$$v_{tg} = \sqrt{\frac{r(e^{\nu})'}{2e^{\nu}}}. \tag{12b}$$

From the equations (8) and (11), Eq. (12a) becomes

$$v_{tg}^2 = \frac{1}{2}\frac{C + Q\frac{1}{3}k^4 r^3(U+2P)}{r\left(1 - \frac{C}{r} - \frac{Q}{r}\right)}. \tag{13}$$

Thought those relations, we see that is possible to determine the Weyl fluid parameter in such a manner that it satisfies the flat rotation curve condition. In order to implement that, we consider the foreword condition [8]

$$e^{\nu} = B_0 r^l, \tag{14}$$

where B is a constant and $l = 2v_0^2$, with v_0 being the asymptotical rotation curve velocity. When the above relation is take into Eq. (12a) we confirm that the tangential velocity take on the constant value v_0.

The constraint imposed by Eq. (14) allows us to find the metric components. Using that in Eqs. (5), (6) and (7) we find

$$e^{-\lambda} = \frac{2}{(2+\frac{l}{2})a} + \frac{D}{r^a}, \tag{15}$$

where D is an integration constant, $l = 2v_0^2$ and $a = \frac{\left(2+l+\frac{l^2}{2}\right)}{(2+\frac{l}{2})}$. Thus, the metric equation is given by

$$ds^2 = -(B_0 r^l)dt^2 + \left[\frac{2}{(2+\frac{l}{2})a} + \frac{D}{r^a}\right]^{-1} dr^2 + r^2 d\Omega^2. \tag{16}$$

Furthermore, the expression for the dark radiation can be obtained by using Eq. (15) into (5)

$$U(r) = k^{-4}\left[\frac{D(a-1)}{r^{a+2}} + \left(1 - \frac{2}{a(2+\frac{l}{2})}\right)\frac{1}{r^2}\right]. \tag{17}$$

The dark pressure term is determined by the above relation together with Eqs. (8) and (14)

$$P(r) = -\frac{d}{(2+a-c)r^{2+a}} - \frac{e}{(2-c)r^2} + \frac{E}{r^c}, \tag{18}$$

where E is a integration constant, $c = \frac{6+l}{2}$, $d = \frac{D(a-1)[a+2(1-l)]}{2k^4}$ and $e = k^{-4}(1-l)\left(1 - \frac{2}{a(2+\frac{l}{2})}\right)$.

4. Dynamical Systems Analysis

In the previous section, we have obtained relations for the Weyl fluid parameters, as well as, for the spacetime metric, that satisfy the flat rotation curve's condition over test particles orbiting the exterior region of spiral galaxies. Let us now analyze whether the solution obtained is stable or not. To implement that, we consider the following transformations

$$q = \frac{C}{r} + \frac{Q}{r}, \tag{19}$$

$$\mu = 3\alpha r^2 U, \tag{20}$$

$$p = 3\alpha r^2 P. \tag{21}$$

When the above transformations are plugged into Eqs. (10) and (11), we get

$$\frac{dq}{d\theta} = \mu - q \tag{22}$$

and

$$\frac{d\mu}{d\theta} = 2\mu - 2p - 2\frac{dp}{d\theta} - \frac{(2\mu+p)[(\mu+2p)/3+q]}{1-q}. \tag{23}$$

The Eqs. (22) and (23) are the so-called structure equations, previously considered in [13, 15]. Here we proceed with a dynamical system analysis to reach a stable vacuum. Nevertheless, there is 3 variables (μ, p and q) to be considered in the structure equations. This fact may be a problem in the dynamical system analysis because it usually leads to a strange attractor [16]. To overcome this problem, let us consider previously an equation of state for the dark fluid parameters. A good choice, for now, is to consider p constant. With this constraint, Eq. (23) becomes

$$\frac{d\mu}{d\theta} = 2\mu - 2p - \frac{(2\mu + p)[(\mu + 2p)/3 + q]}{1 - q}. \tag{24}$$

Therefore, this equation together with Eq. (22) forms an autonomous dynamical system. From (21) we see yet that the p constant condition leads to the relation $P = \beta/r^2$, being β a constant, and we can write $p = 3\alpha\beta$.

In order to determine the fixed points from that dynamical system, we consider initially the condition $\frac{dq}{d\theta} = 0$, that implies $\mu_* = q_*$, where $*$ denotes a fixed point. The condition $\frac{d\mu}{d\theta} = 0$ imposed over (24) gives us the fixed point

$$\mu_* = \frac{3}{14}\left[1 - \alpha\beta \pm \sqrt{1 - 3\alpha\beta(9\alpha\beta + 10)}\right]. \tag{25}$$

The equations (10), (19) and (20) with the condition $\mu_* = q_*$, give

$$r\frac{dQ}{dr} = C + Q. \tag{26}$$

Taking the derivative on both sides of the above equation with respect to r, we get $\frac{d^2Q}{dr^2} = 0$, which implies

$$\frac{dQ}{dr} = 3\alpha r^2 U = \delta, \tag{27}$$

where δ is a constant. Therefore, from the above relation and $P = \beta/r^2$, we get

$$P = \frac{3\alpha\beta}{\delta}U. \tag{28}$$

This relation is nothing but an equation of state for the Weyl fluid parameters. The proportionality between U and P is a condition for the existence of a quasi-homologous symmetry on the brane [13].

To determine the stability of the critical point, it is necessary to analyze the eigenvalues of its Jacobian [17], we obtain

$$\lambda_\pm = \frac{1}{\mu_* - 1}\left[\frac{19\mu_*}{6} + \frac{5\alpha\beta}{2} - \frac{3}{2} \pm \sqrt{193\mu_*^2 + 570\mu_*\alpha\beta - 6\mu_* + 441\alpha^2\beta^2 + 18\alpha\beta + 9}\right]. \tag{29}$$

Let us analyze the behavior of the eigenvalue considering the three possibilities for p.

In the $p = 0$ case, we have $\beta = 0$, and the critical points (25) are $\mu_*^+ = 3/7$ and $\mu_*^- = 0$. Take these points in the equation (29), we get the eigenvalues associated to the critical points. For μ_*^+, we have

$$\lambda_+ = -\frac{1}{8}(-2 + \sqrt{228}) < 0 \tag{30}$$

and

$$\lambda_- = -\frac{1}{8}(-2 - \sqrt{228}) > 0, \tag{31}$$

characterizing a saddle point. We remark parenthetically that the spacetime associated to this critical point is unstable under small perturbations.

For the other critical point ($\mu_*^- = 0$), we get

$$\lambda_- = 2\lambda_+ = 2, \tag{32}$$

which is a repellor.

Now, we consider the $p > 0$ case, leading to $\beta > 0$. From equation (25), we note that

$$\alpha\beta(9\alpha\beta + 10) \leq 1/3, \tag{33}$$

in order to have a real fixed point. Those relations, imply that the range of $\alpha\beta$ factor in equation (25) is within the interval $\left(0, \frac{-30+\sqrt{1008}}{54}\right]$. Taking the equation (25) into (29), for the above range, we conclude that for the fixed point μ_*^+ the eigenvalue λ_- is always positive [18]. Therefore, μ_*^+ is either saddle or repellor. Moreover, for the μ_*^- fixed point, it is always a repellor [18].

Let us consider now the $p < 0$ case. For real fixed point the condition

$$\alpha|\beta|(10 - 9\alpha|\beta|) \geq -\frac{1}{3} \tag{34}$$

holds. Thus the range of $\alpha|\beta|$, for this case, is $\left(0, \frac{30+\sqrt{1008}}{54}\right]$. Replacing the critical points given by the relation (25) into (29), we find for this range, that the critical point μ_*^+ is always a saddle point [18]. On the other hand, the critical point μ_*^- has a singular behavior in the considered range: the eigenvalues at this last critical point are both positive until $p/3 = 1$, and therefore it is classified as a reppelor. But, from $p/3 = 1$ on, both eigenvalues become negative, and the

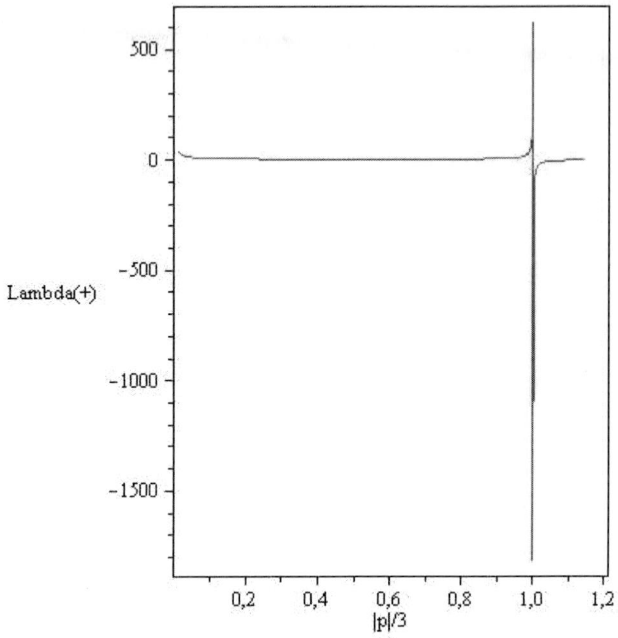

Figure 2. Eigenvalue λ_+ for the fixed point μ_*^-. In this plot, the horizontal axis has been shortened for the sake of exposition [18].

critical point μ_*^- starts to act as an attractor (Figs. (2) and (3)). This behavior is known in dynamical system theory as the so-called transcritical bifurcation.

Therefore, the dynamical system analysis, gives us that the value for the dark pressure, associated with the Weyl fluid, should assume negative values in order to give rise a stable solution.

5. Final Remarks

Despite of the vast majority of works dealing with dark matter hypothesis to explain galaxy rotation curves and other correlated astrophysical problems, we could see here that a scenario with spatial extra dimension can supply an alternative explanation for these problems.

The Weyl-projected correction term, coming from the bulk gravitation,

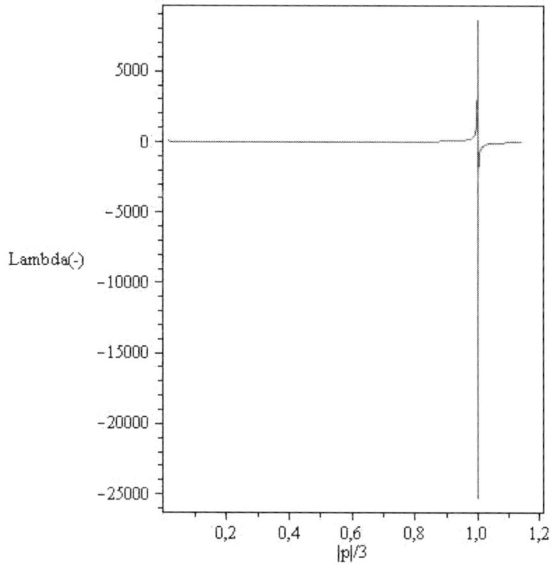

Figure 3. Eigenvalue λ_- for the fixed point μ_*^-. In this plot, the horizontal axis has been shortened for the sake of exposition [18].

modifies the vacuum gravitational field on the brane, acting as a cosmological fluid with associated dark radiation and dark pressure parameters. We see that an appropriate choice for the parameters can describe the flat rotation curves profile in the spiral galaxies.

Throughout the dynamical system analysis, we see that the fixed points give rise to an equation of state already obtained [13] by imposing the quasi-homologous symmetry over the brane vacuum solution. Besides, we have obtained a restriction for the dark pressure parameter: it is necessary a negative value for p in order to have an attractor fixed point.

Moreover, we see that the dark pressure profile $P = -|\beta|/r^2$, coming from the dynamical system analysis, is suitably compatible with the expression for the dark pressure (18), obtained in the Section 2.

Apart of the explanation of galaxy rotation curves, it is noteworthy that brane world scenarios can be used to describe other phenomena usually treated by means of the dark matter hypotheses, as the structure formation in the primordial universe [19] and gravitational lensing by galaxy clusters [20].

Acknowledgments

M. C. B. Abdalla and J. M. Hoff da Silva thanks to Conselho Nacional de Desenvolvimento Científico e Tecnológico (CNPq) for partial financial support (482043/2011-3; 308623/2012-6). P. F. Carlesso thanks to CAPES-Brazil for financial suport.

References

[1] F. Zwicky, *Act. Helv. Phys.* 6, 110 (1933).

[2] A. Bosma, *The distribution and kinematics of neutral hydrogen in spiral galaxies of various morphological types*, (PhD dissertation, The University of Groningen, 1978).

[3] T. S. Van Albada and R. Sancisi, *Phil. Trans. R. Sac. Land.* A 320, 447 (1986).

[4] P. Salucci and G. Gentile, *Phys. Rev.* D 73, 128501 (2006).

[5] T. Shiromizu, K. Maeda, and M. Sasaki, *Phys. Rev.* D 62, 043523 (2000).

[6] M. K. Mak and T. Harko, *Phys. Rev.* D 70, 024010 (2004).

[7] C. G. Bhmer and T. Harko, *Class. Quantum Grav.* 24, 3191 (2007).

[8] F. Rahaman, M. Kalam, A. DeBenedictis, A. A. Usmani e S. Ray, *Monthly Notices of the Royal Astronomical Society* 389, 27 (2008).

[9] K. K. Nandi, A. I. Filippov, F. Rahaman, Saibal Ray, A. A. Usmani, M. Kalam e A. DeBenedictis, *Monthly Notices of the Royal Astronomical Society* 399, 2079 (2009).

[10] L. Randall and R. Sundrum, *Phys. Rev. Lett.* 83, 3370 (1999); L. Randall and R. Sundrum, *Phys. Rev. Lett.* 83, 4690 (1999).

[11] R. Maartens, Reference Frames and Gravitomagnetism, Ed. J. Pascual-Sanchez et al., *World Sci.*, 93 (2001).

[12] S. Chandrasekhar, *The Mathematical Theory of Black Holes*, Oxford University Press (1983).

[13] T. Harko and M. K. Mak, *Phys. Rev. D* 69, 064020 (2004).

[14] C. G. Bohmer and T. Harko, *Class. Quant. Grav.* 24, 3191 (2007).

[15] C. G. Bohmer, T. Harko, and V. Sabau, *Adv. Theor. Math. Phys.* 16, 1145 (2012).

[16] D. Ruelle and F. Takens, *Commun. Math. Phys.* 20, 167 (1971); R. Plykin, *Sitz. Ber. Math.* 23, 333 (1974); S. Newhouse, D. Ruelle, and F. Takens, *Commun. Math. Phys.* 64, 35 (1978).

[17] J. Hale and H. Koçak, *Dynamics and Bifurcations*, Springer-Verlag, New York (1991).

[18] M. C. B. Abdalla, P. F. Carlesso and J. M. Hoff da Silva, *Eur. Phys. J. C* 73, 2452 (2013).

[19] S. Pal, *Phys. Rev.* D 74, 024005 (2006).

[20] T. Harko e K. S. Cheng, *The Astrophysical Journal* 636, 8 (2006).

In: Research Development in Dark Matter Research ISBN: 978-1-62948-010-7
Editors: Nori Kinjo and Akira Nakajima

Chapter 8

FROM MASSIVE NEUTRON STARS TO DARK MATTER PARTICLE PROPERTIES

Ang Li[1,*]***, Feng Huang***[1,†] ***and Ren-Xin Xu***[2,‡]
[1]Department of Astronomy and Institute of Theoretical Physics and Astrophysics, Xiamen University, Xiamen, China
[2]Physics and State Key Laboratory of Nuclear Physics and Technology, Peking University, Beijing, China

Abstract

Newly-determined mass of $1.97 \pm 0.04\ M_\odot$ for PSR J1614-2230 has been a challenge for the neutron star with a hyperon core (namely hyperon star), since hyperons usually reduce the theoretical maximum mass of the star. In this article, we consider dark matter as another possible constituent in hyperon stars' interior to loose this mass constrain. We take dark matter as self-interacting Fermi gas with certain repulsive interaction among the dark matter particles and assume non-interaction between dark matter and ordinary matter in the neutron star system. We find that the star maximum mass is sensitive to the particle mass of dark matter, and a high enough star mass larger than $2\ M_\odot$ could be achieved when the particle mass is small enough. In this particular model, a strong upper limit 0.64

*E-mail address: liang@xmu.edu.cn
†E-mail address: fenghuang@xmu.edu.cn
‡E-mail address: r.x.xu@pku.edu.cn

GeV for dark matter mass is obtained in strongly-interacting dark matter and 0.16 GeV for dark matter mass in weakly-interacting dark matter. The latter is commonly considered it is the case for typical dark matter particles. Dark matter accumulated around the star could also contribute to the mass measurement, however, such contribution could be safely ignored when the generally used dilute dark matter density is assumed.

PACS: 26.60.Dd, 26.60.Kp, 95.35.+d, 97.60.Jd

Keywords: Neutron stars; Equation of state; Dark matter

1. Introduction

The recent measurement of the Shapiro delay in the radio pulsar PSR J1614-2230 yielded a mass of $1.97 \pm 0.04\ M_\odot$ [1]. Such a high neutron star (NS) mass might rule out many predictions of non-nucleonic components (free quarks, mesons, hyperons) in NS interiors, since they usually reduce the theoretical maximum mass of the star [2, 3, 4]. For example, a large NS maximum mass larger than 2 $M_\odot$ is obtained from nucleonic equation of state (EoS) from the microscopic Brueckner theory, but a rather low value below 1.4 $M_\odot$ is found for hyperon stars (HSs) in the same method [5, 6]. It is the so-called hyperon puzzle. Although the present calculation did not include three-body hyperon interaction due to the complete lack of experimental and theoretical information, it seems difficult to imagine that these could strongly increase the maximum mass, since the importance of hyperon-hyperon potentials should be minor as long as the hyperonic partial densities remain limited. However, if there is universal strong repulsion in all relevant channels the maximum mass may be significantly raised [7], so the including of an improved hyperon-nucleon and hyperon-hyperon potentials and hyperonic three-body force is still appealing to settle this apparent contradiction, which needs further experimental data. In addition, the presence of a strongly-interacting quark matter, in the star's interior (i.e., hybrid star), is proposed to be a good candidate for troubleshooting this problem [8], However, NS masses substantially above 2 $M_\odot$ seem to be out of reach even for hybrid stars using most of effective quark matter EoS (bag model [9], NJL model [10], color dielectric model [11]), A hybrid star with 2 $M_\odot$ is only allowed when using the Dyson-Schwinger approach for the description of quark matter [12].

Alternatively, dark matter (DM), as another possible constituent in NS interior, has been taken into account and a new kind of compact star, i.e., DM-admixed NS, has been studied in several articles [13, 14, 15, 16, 17, 18, 19, 20, 21]. The effect of the DM accreted by the NS is especially important for NSs because of their sufficiently compact mass which could do the accretion efficiently. The general effect induced by DM inside NS is complicated due to the lack of information about the particle nature of DM. DM could annihilate, such as Majorana self-annihilating particles, which may lead to sizable energy deposit and then enhance the thermal conductivity or trigger the deconfinement phase transition in the core of NS for the emergency of a quark star, as illustrated by Perez-Garcia et al. in [13]. Such quark star objects are at present very uncertain in theory and could easily accord with astrophysical measurements within the modification of model parameters [22, 23]. Another generally considered DM candidate is the non-self-annihilating particle, such as the newly interesting mirror DM ([24] and references therein) or asymmetric DM ([25] and references therein). When they accumulate in NSs, the resulting maximum mass is then rather sensitive to the EoS model of DM. Assuming that the DM component is governed by an ideal Fermi gas, Leung et al. [14] studied the various structures of the DM-admixed NSs by solving the relativistic two-fluid formalism. Ciarcelluti & Sandin [15] approximated the EoS of mirror matter with that of ordinary nuclear matter, varied the relative size of the DM core, and explained all astrophysical mass measurements based on one nuclear matter EoS.

In this chapter, we will consider non-self-annihilating DM particles as fermions, and the repulsive interaction strength among the DM particles is assumed to be a free parameter $m_{\mathcal{I}}$ as in [26]. Different to previous DM-admixed NS models, we take the total pressure (energy density) as the simple sum of the DM pressure (energy density) and NS pressure (energy density), the general dependence of the mass limit on DM particle mass and the interaction strength is then presented based on the present model. Furthermore, in the particular case of PSR J1614-2230, we notice that the DM around the star should also contribute to its mass measurement due to the pure gravitational effect. However, our numerically calculation illustrates that such contribution could be safely ignored because of the usual diluted dark matter DM environment assumed. In the early universe, DM and normal matter also interact with each other. However, the interaction between DM and baryonic matter is model-dependent, and for the sake of simplicity, we would ignore this at first in the present chapter to see the primary role of DM, and may do further calculations in the future.

The chapter is arranged as follows. The details of our theoretical model are presented in §2, followed by the numerical results. Out conclusion is given in §3.

2. Model

We begin with the description of our model of DM-admixed NS, which mainly deals with the DM particles scattered inside the star. Those DM particles would modify the local pressure-energy density relationship of the matter and hence change the theoretical prediction of the gravitational mass of the star. The structure equations for static compact stars, the Tolman-Oppenheimer-Volkov equations in the non-rotating case, are written as:

$$\frac{dP(r)}{dr} = -\frac{Gm(r)\mathcal{E}(r)}{r^2}\frac{\left[1+\frac{P(r)}{\mathcal{E}(r)}\right]\left[1+\frac{4\pi r^3 P(r)}{m(r)}\right]}{1-\frac{2Gm(r)}{r}}, \tag{1}$$

$$\frac{dm(r)}{dr} = 4\pi r^2 \mathcal{E}(r) \tag{2}$$

being G the gravitational constant. P and $\mathcal{E}$ denote the pressure and energy density. Here we have neglected the rotating of the stars. A recent study by Perez-Garcia & Joseph Silk [27] has pointed out that DM accretion could give rise to a change in the rotational pattern. The EoS of the star, relating P and $\mathcal{E}$, is needed to solve the above set of equations. In our DM-admixed NS model, $P = P_N + P_\chi, \mathcal{E} = \mathcal{E}_N + \mathcal{E}_\chi$, with the subscript $N(\chi)$ representing NS matter (DM).

The EoS of the ordinary NS matter is handled in the following way: (i) We treat the interior of the stars as β-equilibrium hypernuclear matter, with certain amount of leptons to maintain charge neutrality. The hadronic energy density we use in the article is based on the microscopic parameter-free Brueckner-Hartree-Fock nuclear many-body approach, employing the latest derivation of nucleon-nucleon microscopic three-body force [28], supplemented by the very recent Nijmegen extended soft-core ESC08b hyperon-nucleon potentials [8]. The EoS can be computed straightforwardly after adding the contributions of the noninteracting leptons [8]. (ii) For the description of the NS crust, we join the hadronic EoS with those by Negele and Vautherin [29] in the medium-density regime, and those by Feynman-Metropolis-Teller [30] and

Baym-Pethick-Sutherland [31] for the outer crust. No temperature effect is considered in the present work and the NS system is assumed to be cold.

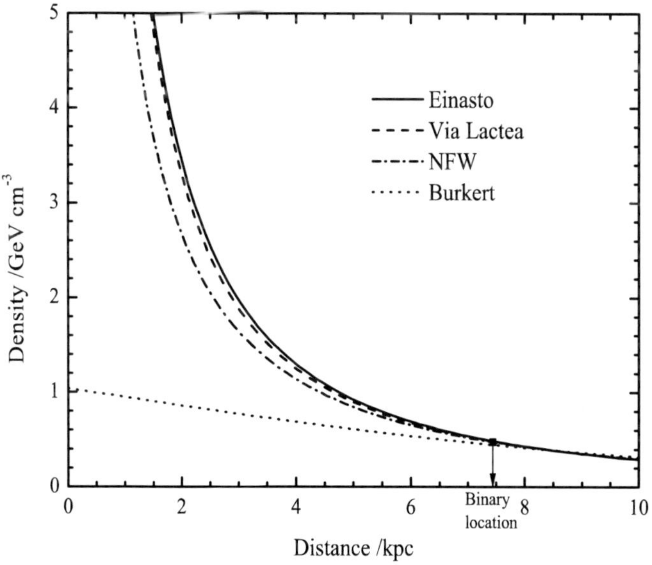

Figure 1. Four models of the galactic DM density profiles employed in the paper. *Einasto* stands for the best-fitting Einasto density profile obtained from the results of the Aquarius simulation [33]. *Via Lactea* stands for a profile extrapolated from the Via Lactea II simulation [33]. *NFW* stands for the prototypical Navarro-Frenk-White density profile [34]. *Burkert* stands for the Burker profile [35, 36]. The location of the binary system r = 7.44 kpc is indicated with an arrow. Taken from Ref. [37].

The EoS of DM is modeled to be that of a self-interacting Fermi gas with one parameter $m_{\mathcal{I}}$ accounting for the energy scale of the interaction [26]. For weak interaction (WI) the scale $m_{\mathcal{I}}$ is just given by Fermi's constant or the vacuum expectation of the Higgs field generating the masses of the W and Z bosons, which is around $\sim$ 300 GeV [26]. For strongly interacting (SI) DM particles, $m_{\mathcal{I}}$ is assumed to be $\sim$ 100 MeV [26], as a typical interaction mass scale according to the gauge theory of the strong interactions. This is a wide enough range of energy scale, and we hope the calculation would cover most of the promising DM candidates.

Then we consider the mass contribution of DM halo via gravitational capture. We estimate the size of the halo R as big as that of the possible Roche lobe [32] of the centered PSR J1614-2230. Based on its calculated distance from the galactic center, we take an average value for the DM density, $\bar{\rho_\chi}$ = 0.474 GeV/cm^3, from evaluating several spherically symmetric Galactic DM profiles [33, 34, 35, 36] (as shown in Fig. 1). Then we get the contributed mass of gravitationally captured DM particles as $M_\chi = \frac{4}{3}\pi R^3 \bar{\rho_\chi}$ where the size of the star ($\sim$ 10 km) has been neglected compared to its large Roche lobe ($\sim 10^6$ km), and the halo has been regarded as an ideal spherical object.

Our calculated result is presented in Fig. 2 with a shallowed area indicating the mass limit of DM particle m_χ, assuming the observed PSR J1614-2230 is a HS. The $\sim$ 2 $M_\odot$ limit is indicated with a horizontal line. The upper line corresponds to $m_\mathcal{I}$ = 100 MeV (SI case), and the lower line to $m_\mathcal{I}$ = 300 GeV (WI case). Specially we mention that the extra mass measurement contribution from the extended halo is only around 10^{-24} $M_\odot$, which could be safely ignored for the analysis of Shapiro delay. This mass limit from the compact star can be referred as an upper limit for the mass of non-self-annihilating DM particles, namely, it should obey $m_\chi < 0.64$ GeV for SI DM and < 0.16 GeV for WI DM.

3. Conclusion

In summary, to solve the recent hyperon puzzle provoked by the recent 2-solar-mass NS measurement, we present a consistent DM-admixed NS model to investigate the possible influence of DM on the NS mass measurement, and find a strong upper limit for the particle mass of DM from the recent 2 $M_\odot$ NS mass measurement. We take DM as Fermi-gas with certain repulsive interaction among the DM particles and non-interaction between DM and ordinary matter as is generally assumed. As a result, for a fixed DM particle mass, the including of the DM ingredient softens the total EoS and results in a reduced maximum mass. However, the final results are sensitive to the adopted DM particle mass. The smaller the DM particle mass, the harder the EoS or the larger the maximum mass. The observed very massive NS requires a very stiff EoS and then sets a strong upper limit on the DM particle mass. In our numerical calculation, DM particle mass should less than 0.64 GeV for SI DM and 0.16 GeV for WI DM. In order to relax such strong constraint, we further consider the possible extended DM halo contribution to the particular mass measurement. However,

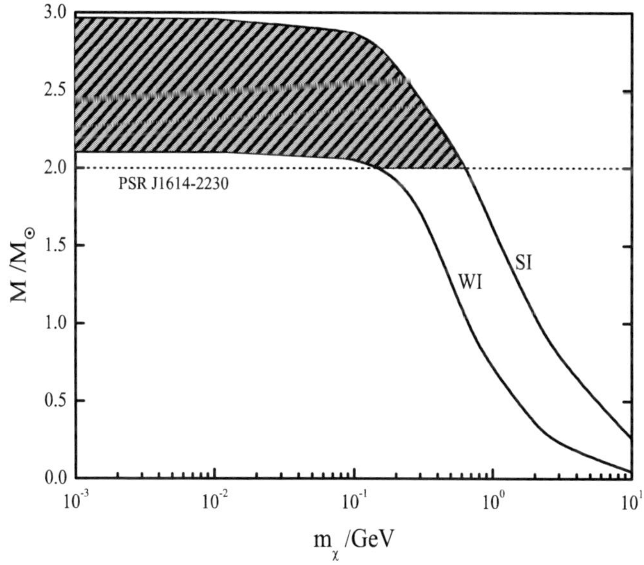

Figure 2. HSs' maximum masses as a function of the particle mass of DM candidates m_χ. The upper line is for $m_\mathcal{I}$ = 100 MeV (SI case), and the lower line for $m_\mathcal{I}$ = 300 GeV (WI case). The $\sim$ 2 $M_\odot$ limit of PSR J1614-2230 is indicated with a horizontal line. Taken from Ref. [37].

due to the diluted DM environment, such kind of contribution could be safely ignored.

By incorporating the possible influence of dark matter particles accumulated into and around neutron stars, we study the influence of dark matter on the mass measurement of neutron stars. We take dark matter as self-interacting Fermi-gas with certain repulsive interaction among the dark matter particles and non-interaction between dark matter and ordinary matter as is generally assumed. The mass contribution from the extended dark matter halo is found to be ignorable. However, the neutron star maximum mass is sensitive to the particle mass of dark matter, and a high enough neutron star mass larger than 2 $M_\odot$, newly-determined for PSR J1614-2230, could be achieved when the particle mass is small enough, being $M_\odot$ the mass of the sun. Finally a strong upper limit (0.64 GeV) is obtained for the particle mass of non-self-annihilating dark matter from the constrain of the recent 2 $M_\odot$ mass limit for neutron stars.

Acknowledgments

We would like to thank Prof. M. Angeles Perez-Garcia for reading carefully our contribution and showing many medicinal comments and suggestions. This chapter is funded by the National Basic Research Program of China (Grant No 2009CB824800), the National Natural Science Foundation of China (Grant No 10905048, 10973002, 10935001, 11078015), and the John Templeton Foundation.

References

[1] Demorest, PB; Pennucci, T; Ransom, SM; Roberts, MSE; Hessels, JWT. A two-solar-mass neutron star measured using Shapiro delay. *Nature*, (2010) 467, 1081-1083.

[2] Heiselberg, H; Hjorth-Jensen, M. Phases of dense matter in neuron stars. *Phys. Rep.*, (2000) 328, 237-327.

[3] Schaffner-Bielich, J. Strange quark matter in stars: a general overview. *J. Phys. G : Nucl. Part. Phys.*, (2005) 31, S651-S657.

[4] Li, A; Zhou, XR; Burgio, GF; Schulze, HJ. Protoneutron stars in the Brueckner-Hartree-Fock approach and finite-temperature kaon condensation. *Phys. Rev. C*, (2010) 81, 025806.

[5] Burgio, GF; Schulze, HJ. The maximum and minimum mass of protoneutron stars in the Brueckner theory. *Astron. Astrophys.*, (2010) 518, A17.

[6] Burgio, GF; Schulze, HJ; Li, A. Hyperon stars at finite temperature in the Brueckner theory. *Phys. Rev. C*, (2011) 83, 025804.

[7] Vidaña, I; Logoteta, D; Providencia, C; Polls, A; Bombaci, I. Estimation of the effect of hyperonic three-body forces on the maximum mass of neutron stars. *Europhys. Lett.*, (2011) 94, 11002.

[8] Schulze, HJ; Rijken, T. Maximum mass of hyperon stars with the Nijmegen ESC08 model. *Phys. Rev. C*, (2011) 84, 035801.

[9] Burgio, GF; Baldo, M; Sahu, PK; Schulze, HJ. Hadron-quark phase transition in dense matter and neutron stars. *Phys. Rev. C*, (2002) 66, 025802.

[10] Baldo, M; Buballa, M; Burgio, GF; Neumann, F; Oertel, M; Schulze, HJ. Neutron stars and the transition to color superconducting quark matter. *Phys. Lett. B*, (2003) 562, 153-160.

[11] Maieron, C; Baldo, M; Burgio, GF; Schulze, HJ. Hybrid stars with the color dielectric and the MIT bag models. *Phys. Rev. D*, (2004) 70, 043010.

[12] Chen, H; Baldo, M; Burgio, GF; Schulze, HJ, Hybrid stars with the Dyson-Schwinger quark model. *Phys. Rev. D*, (2011) 84, 105023.

[13] Perez-Garcia, MA; Silk, J; Stone, JR. Dark Matter, Neutron Stars, and Strange Quark Matter. *Phys. Rev. Lett.*, (2010) 105, 141101.

[14] Leung, SC; Chu, MC; Lin, LM. Dark-matter admixed neutron stars. *Phys. Rev. D*, (2011) 84, 107301.

[15] Ciarcelluti, P; Sandin, F. Have neutron stars a dark matter core? *Phys. Lett. B*, (2011) 695, 19-21.

[16] Goldman, I. Implications of Mirror Dark Matter on Neutron Stars. *Acta Phys. Polon. B*, (2011) 42, 2203-2209.

[17] Sandin, F; Ciarcelluti, P. Effects of mirror dark matter on neutron stars. *Astroparticle Phys.*, (2009) 32, 278-284.

[18] Bertone, G; Fairbairn, M. Compact stars as dark matter probes. *Phys. Rev. D*, (2008) 77, 043515.

[19] Kouvaris, C. WIMP annihilation and cooling of neutron stars. *Phys. Rev. D*, (2008) 77, 023006.

[20] Kouvaris, C; Tinyakov, P. Can neutron stars constrain dark matter? *Phys. Rev. D*, (2010) 82, 063531.

[21] de Lavallaz, A; Fairbairn, M. Neutron stars as dark matter probes. *Phys. Rev. D*, (2010) 81, 123521.

[22] Li, A; Xu, RX; Lu, JF. Strange stars with different quark mass scalings. *Mon. Not. R. Astron. Soc.*, (2010) 402, 2715-2719.

[23] Li, A; Peng, GX; Lu, JF. Strange star candidates revised within a quark model with chiral mass scaling. *RAA*, (2011) 11, 482-490.

[24] Okun, L. B. Mirror particles and mirror matter: 50 years of speculations and search, *Phys. Usp.*, (2007) 50, 380-389.

[25] An, HP; Chen, SL; Mohapatra, RN; Zhang, Y. Leptogenesis as a common origin for matter and dark matter. *J. High Energ. Phys.*, (2010) 3, 124.

[26] Narain, G; Schaffner-Bielich, J; Mishustin, IN. Compact stars made of fermionic dark matter. *Phys. Rev. D*, (2006) 74, 063003.

[27] Perez-Garcia, MA; Silk, J. Dark matter seeding and the kinematics and rotation of neutron stars. *Phys. Lett. B*, (2012) 711, 6-9

[28] Li, ZH; Lombardo, U; Schulze, HJ; Zuo, W. Consistent nucleon-nucleon potentials and three-body forces. *Phys. Rev. C*, (2008) 77, 034316.

[29] Negele, JW; Vautherin, D. Neutron star matter at sub-nuclear densities. *Nucl. Phys. A*, (1973) 207, 298-320.

[30] Feynman, RP; Metropolis, N; Teller, E. Equations of state of elements based on the generalized Fermi-Thomas theory. *Phys. Rev. C*, (1949) 75, 1561-1573.

[31] Baym, G; Pethick, C; Sutherland, P. The ground state of matter at high densities: equation of state and stellar models. *Astrophys. J.*, (1971) 170, 299-317.

[32] Eggleton, PP. Approximations to the radii of Roche lobes. *Astrophys. J.*, (1983) 268, 368-369.

[33] Springel, V; Wang, J; Vogelsberger, M; Ludlow, A; Jenkins, A; Helmi, A; Navarro, JF; Frenk, CS; White, SDM. The Aquarius Project: the subhaloes of galactic haloes. *Mon. Not. R. Astron. Soc.*, (2008) 391, 1685-1711.

[34] Navarro, JF; Frenk, CS; White, SDM. A Universal Density Profile from Hierarchical Clustering. *Astrophys. J.*, (1997) 490, 493.

[35] Burkert, A. The Structure of Dark Matter Halos in Dwarf Galaxies. *Astrophys. J.*, (1995) 447, L25.

[36] Salucci, P; Burkert, A. Dark Matter Scaling Relations. *Astrophys. J.*, (2000) 537, L9.

[37] Li, A; Huang, F; Xu, RX. Too massive neutron stars: The role of dark matter? *Astroparticle Phys.*, (2012) 37, 70-74.

In: Research Development in Dark Matter Research ISBN: 978-1-62948-010-7
Editors: Nori Kinjo and Akira Nakajima

Chapter 9

MODELING DARK MATTER HALOS VIA SOLITONIC BE CONDENSATES

David Castañeda Valle* and Eckehard W. Mielke†
Departamento de Física,
Universidad Autónoma Metropolitana–Iztapalapa,
México, D.F., Mexico

Abstract

Instead of fluid type dark matter (DM) models, axion-like scalar fields with a periodic self-interaction are analyzed as a model of galaxy halos. Such a cold *Bose-Einstein* type condensate can provide a viable model for a *soliton type* interpretation of the DM 'bullets' observed by means of gravitational lensing in merging galaxy clusters. We study solitary waves for several self-interaction potentials in the relativistic Klein-Gordon equation, mainly in lower dimensions, and comment on their prospective shape-invariant collisions.

Keywords: Dark matter fluids, universal axion, scalar field models, dark matter halos, galaxy clusters

PACS: 95.35.+d, 98.35Gi

*E-mail address: casvada@gmail.com
†E-mail address: ekke@xanum.uam.mx

1. Introduction

Solitary waves have been studied since the legendary discovery in a narrow water channel by John Scott Russell in August of 1834. However, solutions of the *soliton* type have been ignored for quite some time until in 1965 Zabusky and Kruskal [26] could describe its mathematical structure and corroborated the reconstruction of the initial state during non-linear interactions. It turned out that every soliton is a localized solitary wave, but not all localized waves behave like solitons being exactly shape-invariant after collisions.

Nowadays, there exists a host of non-linear differential equations exhibiting soliton type solutions: Well-known are the equations of Korteweg–de Vries (KdV) in two dimensions (2D), those of Kadomtsev–Petviashvili (KP) which supply solitons in three dimensions, the nonlinear Schrödinger (NLS) equation, that of Boussinesq, the nonlinear Klein–Gordon (NLKG) equation, and the sine-Gordon (SG) equation, e.g. The latter has been used in order to model various phenomena, for example, in crystal dislocations, splay waves in nematic liquid crystals, transmission lines, etc.; cf. [9, 23, 4, 11, 1, 12]. Here we will study it in the context of axion models of dark matter.

From a modern perspective, solitary waves occur in various branches of science, for instance in hydrodynamics, including giant water waves (tsunamis), shock waves in the earth crust studied in seismology, in protein chains, nonlinear optics, plasmas, as well as vortices occurring in the Great Red Spot of Jupiter [25].

In astrophysics they have been analyzed in order to describe compact objects like boson stars (BS). The latter are Bose-Einstein condensations (BEC) on a macroscopic scale, where the self-generated gravitational fields provides a ‘trap’ for the hypothetical scalar fields, cf. Ref. [19, 24]. In particular, solitons built from pseudo-scalars like the axion of quantum chromodynamics (QCD) could serve as a model [20, 6, 18] for dark matter halos.

In various attempts, one would like to describe not only galaxy halos but also their interaction in their collisions million years ago, in order to understand more about the evolution of whole Universe. Famous examples are the *Bullet cluster* [8] as well as the cluster Abell 520 [15]. Here we will probe, if their behavior can be explained, at least partially, by collisions of solitons, though for different regimes of merging velocities, cf. Ref. [7, 5].

2. Non-linear Klein-Gordon Equation

Let us concentrate here on the relativistically invariant Klein-Gordon equation of a single scalar field $\phi = \phi(t, \vec{x})$ with a self-interaction potential $V(\phi)$:

The standard Euler-Lagrange formalism leads to the Klein-Gordon equation

$$\Box\phi = \frac{\partial V(\phi)}{\partial \phi}, \tag{1}$$

where $\Box := \nabla \bullet \nabla - \partial^2/c^2\partial t^2$ is the wave operator of Jean le Rond d'Alembert. Quite generally, this equation turn out to be nonlinear: In a series expansion of the self-interacting potential V, the quadratic term returns the mass term in the KG equation whereas higher order terms account for a possible non-linearity.

Such equations are invariant under the discrete symmetries P and T of parity and time-reflection. Moreover, they are invariant under the standard Lorentz transformations such that one can resolve the equation in a rest frame where a static solution $\phi = \phi(\vec{x})$ only depends on the spatial coordinates. After an inverse Lorentz transformation $\vec{x} \to \gamma(\vec{x} - \vec{u}t)$, where $\gamma = \left(1 - \vec{u} \bullet \vec{u}/c^2\right)^{-1/2}$ is the Lorentz factor and $\vec{u}$ the (phase) velocity of the wave, one can recover a stationary solution in inertial frame moving with relative velocity $\vec{u}$.

Superpotential and Bogomolny Equation in (1+1)D

Let us study two-dimensional toy models and first consider static, i.e. time-independent solution of a non-linear KG equation. Then Eq. (1) reduces to

$$\frac{\partial^2\phi}{\partial x^2} = \frac{\partial V(\phi)}{\partial \phi}. \tag{2}$$

After multiplying both sides by $2\phi'$, the first integral turns out to be the Bogomolny [16] equation

$$\frac{\partial \phi}{\partial x} = \pm\sqrt{2V(\phi)} \tag{3}$$

of first differential order. In the case that the potential $V(\phi)$ arises [16, 3, 2], from a so-called *superpotential* $W(\phi)$ implicitly defined by

$$W_\phi^2 := \left(\frac{dW}{d\phi}\right)^2 = 2V(\phi), \tag{4}$$

the Bogomolny equation can be rewritten as follows

$$\frac{d\phi}{dx} = W_\phi \,, \qquad \frac{d\phi}{dx} = -W_\phi \,. \tag{5}$$

Since the fields are assumed to be static in the rest frame, i.e. $\dot{\phi} = 0$, the rest energy of the soliton simply follows by a further integration as

$$E_0 : \;= \int \left[\frac{1}{2} \left(\frac{d\phi}{dx} \right)^2 + V(\phi) \right] dx \tag{6}$$

$$= \int \left(\frac{dW}{d\phi} \right) \left(\frac{d\phi}{dx} \right) dx \tag{7}$$

$$= \left| W(\phi_+) - W(\phi_-) \right| \,, \tag{8}$$

Here $\phi_\pm = \phi\,(x = \pm\infty)$ are the respective boundary values at infinity.

If solitons connect two minima, a topological sector is fixed, so one can define a *topological charge* Q via integrating the temporal component of the current

$$J^\mu := \frac{1}{L} \epsilon^{\mu\nu} \partial_\nu \phi, \qquad \partial_\mu J^\mu \equiv 0 \tag{9}$$

being conserved due to the Poincaré lemma for smooth fields, where L is the topological width.

3. Axion Potential

In QCD, the axion of inertial mass m is self-interacting via the effective potential

$$V(\phi) = \frac{m^4}{\lambda} \left[1 - \cos\left(\frac{\sqrt{\lambda}}{m} \phi \right) \right] \simeq \frac{m^2}{2} \phi^2 - \frac{\lambda}{4!} \phi^4 + \cdots \tag{10}$$

which implies in 2D the sine-Gordon equation

$$\Box\phi + \frac{m^3}{\sqrt{\lambda}} \sin\left(\frac{\sqrt{\lambda}}{m} \phi \right) = 0 \,. \tag{11}$$

Rescaling variables [23] via

$$\tilde{x} = mx \,, \quad \tilde{t} = mt \,, \quad \tilde{\phi} = \frac{\sqrt{\lambda}}{m} \phi, \tag{12}$$

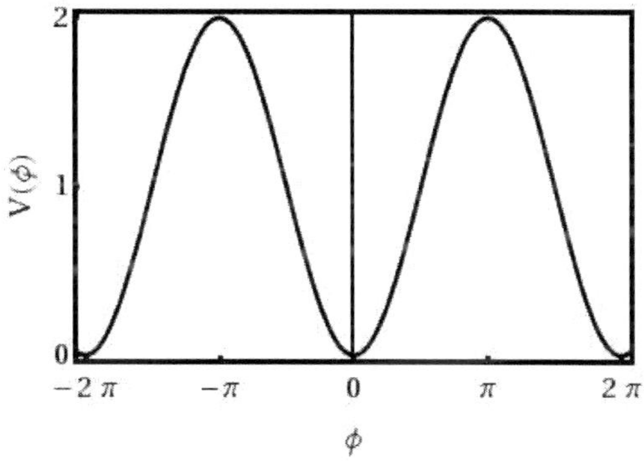

Figure 1. Periodic potential $\tilde{V}\left(\tilde{\phi}\right) = \left(1 - \cos\tilde{\phi}\right)$.

allows us to rewrite the sG equation as

$$\tilde{\Box}\tilde{\phi} + \sin\tilde{\phi} = 0\,. \tag{13}$$

This equation is invariant under $\tilde{\phi}\left(\tilde{x}, \tilde{t}\right) \rightarrow -\tilde{\phi}\left(\tilde{x}, \tilde{t}\right)$ which can be associated with charge conjugation C. The potential has zeros and critical points defined by the conditions

$$V\left(\tilde{\phi}\right) = \frac{m^4}{\lambda}\left(1 - \cos\tilde{\phi}\right) \quad = \quad 0 \tag{14}$$

$$\frac{d}{d\tilde{\phi}}V\left(\tilde{\phi}\right) = \frac{m^4}{\lambda}\sin\tilde{\phi} \quad = \quad 0. \tag{15}$$

Thus, there exist infinite many critical points at $\overline{\tilde{\phi}_n} = 2n\pi$, with equal spacing due to $n \in \mathbb{Z}$. In this case, solutions can interpolate critical points, and in such a topological sector of width 2π, the boundary conditions are $\tilde{x} \rightarrow -\infty$ y $\tilde{x} \rightarrow +\infty$. In view of Eq. (9) the topological charge is simply given by

$$Q = \int J^0 d\tilde{x} = \frac{1}{2\pi}\left[\tilde{\phi}_+ - \tilde{\phi}_-\right]\,. \tag{16}$$

The shape of solitons of the sG equation can be obtain by solving the static equation

$$\frac{\partial^2\tilde{\phi}}{\partial\tilde{x}^2} = \sin\tilde{\phi}\,, \qquad \mathrm{mod}\ 2\pi\,. \tag{17}$$

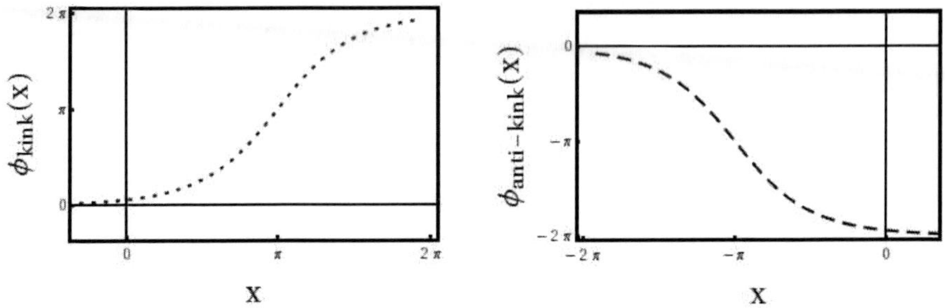

Figure 2. *Kink* solution centered at $x_0 = \pi$, *antikink* solution centered at $x_0 = -\pi$.

Since only two minima with topological charge $Q = \pm 1$ can be connected, the resulting solitons are *kink* y *antikink* type, corresponding to positive and negative charge, respectively.

The periodicity of the axion potential implies the existence of solutions centered at different x_0's, depending on the critical points considered, each giving rise to a *kink* or *antikink*. Then, the Bogomolny equation reads

$$\frac{d\tilde{\phi}}{d\tilde{x}} = \pm\frac{m}{\sqrt{\lambda}}\sqrt{2\left(1-\cos\tilde{\phi}\right)} = \pm 2\frac{m}{\sqrt{\lambda}}\sin\frac{\tilde{\phi}}{2}. \tag{18}$$

Its implicit solution

$$\tilde{x} - \tilde{x}_0 = \pm\frac{\sqrt{\lambda}}{m}\ln\left|\tan\frac{\tilde{\phi}\left(\tilde{x}\right)}{4}\right| \tag{19}$$

can be inverted for the axion type field

$$\tilde{\phi}\left(\tilde{x}\right) = 4\arctan\left\{\exp\left[\pm\frac{m}{\sqrt{\lambda}}\left(\tilde{x}-\tilde{x}_0\right)\right]\right\}. \tag{20}$$

These solutions (Fig. 2) are of *kink* and *antikink* type, The positive or negative sign ($\pm$), respectively, are associated with the topological charge and determine the chirality with which the waves propagate [9, 23]. One identifies the positive sign with a soliton or kink (Fig. 2a), and the negative sign with an antisoliton or antikink (Fig. 2b), in accordance with the symmetry of charge conjugation C of the field within adjacent sectors of the topology.

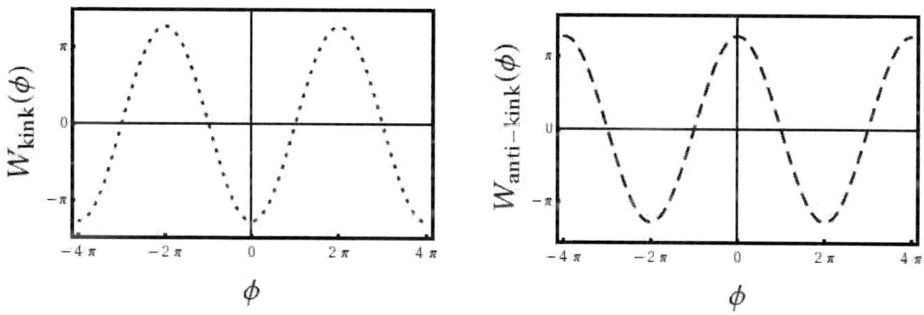

Figure 3. The superpotentials of a kink and antikink, respectively.

In the case of solitons of the kink type, the critical values increment from $\bar{\tilde{\phi}} = 2\pi N$ to $\bar{\tilde{\phi}} = 2\pi(N+1)$ resulting in the charge $Q = +1$; while for antikinks, the values reduce from $\bar{\tilde{\phi}} = 2\pi N$ to $\bar{\tilde{\phi}} = 2\pi(N-1)$ leading to the charge $Q = -1$.

When applying a Lorentz transformation to Eq. (20), we obtain the stationary solution

$$\tilde{\phi}\left(\tilde{x},\tilde{t}\right) = 4\arctan\left[\exp\left(\pm\gamma\frac{m}{\sqrt{\lambda}}\left(\tilde{x}-\tilde{x}_0-u\tilde{t}\right)\right)\right]. \tag{21}$$

An integration leads to the defining equation

$$\frac{dW}{d\tilde{\phi}} = \pm 2\frac{m^3}{\sqrt{\lambda}}\sin\frac{\tilde{\phi}}{2}, \tag{22}$$

for the superpotential. It turns out that

$$W_{\rm k}\left(\tilde{\phi}\right) = -4\frac{m^3}{\sqrt{\lambda}}\cos\frac{\tilde{\phi}}{2} \tag{23}$$

$$W_{\bar{\rm k}}\left(\tilde{\phi}\right) = 4\frac{m^3}{\sqrt{\lambda}}\cos\frac{\tilde{\phi}}{2} \tag{24}$$

like the self-interaction potential is periodic, cf. Fig. 3.

These superpotentials (Fig. 4) facilitate the calculation of the energy of a *kink* as well as of an *antikink* :

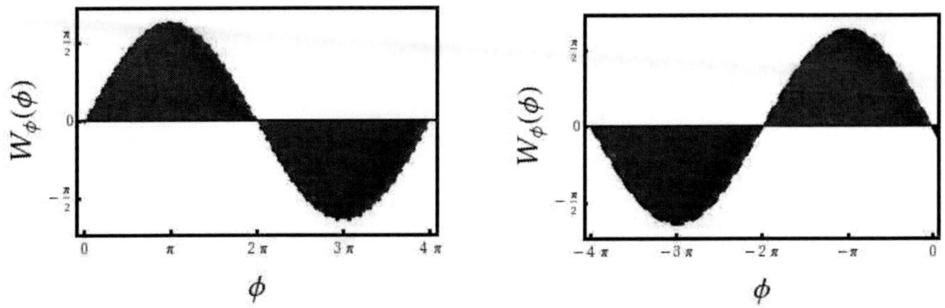

Figure 4. Energy from the superpotential of a *kink* and an antikink, respectively.

$$
\begin{aligned}
E_0 &= \left| W_{\mathrm{k}}\left(\bar{\tilde{\phi}}_n(-\infty)\right) - W_{\mathrm{k}}\left(\bar{\tilde{\phi}}_{n+1}(\infty)\right)\right| c^2 \\
&= \left| W_{\overline{\mathrm{k}}}\left(\bar{\tilde{\phi}}_{n+1}(-\infty)\right) - W_{\overline{\mathrm{k}}}\left(\bar{\tilde{\phi}}_n(\infty)\right)\right| c^2 \\
&= \left| W(2n\pi) - W(2(n+1)\pi)\right| c^2 = 8\frac{m^3}{\sqrt{\lambda}}c^2 .
\end{aligned} \tag{25}
$$

Since the rest energy is $E_{0\mathrm{k}} = m_{\mathrm{k}}c^2$, we can identify the *effective* mass of the kink as $m_{\mathrm{k}} = 8m^3/\lambda$. Applying the Lorentz factor γ, the relativistic energy of a moving kink is, according to Einstein, given by $E = \gamma m_{\mathrm{k}}c^2$ and its momentum by $p_{\mathrm{k}} = \gamma m_{\mathrm{k}}u$.

4. Higgs Type Potential

In the standard model of elementary particles, the inertial masses are generated via symmetry breaking from a quartic Higgs potential like

$$
V(\phi) = \frac{1}{2}\lambda\phi^2\left(\frac{m^2}{\lambda} - \phi^2\right) = V(-\phi) \tag{26}
$$

which is symmetric against reflection of the scalar fields, see Fig. 5.

The static Klein-Gordon equation is

$$
\frac{\partial^2\phi}{\partial x^2} = 2\lambda\phi\left(\frac{m^2}{2\lambda} - \phi^2\right), \tag{27}
$$

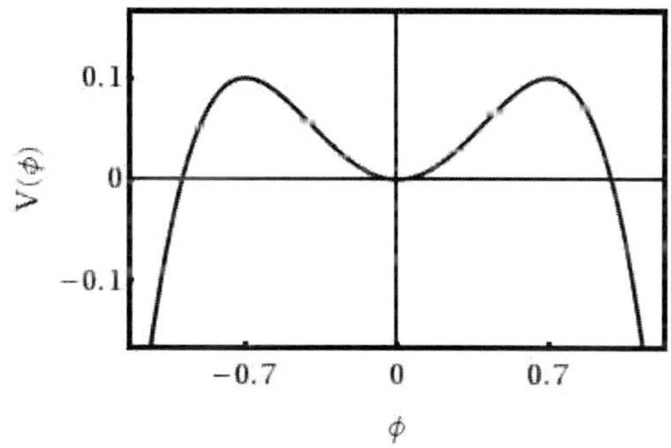

Figure 5. Inverted Mexican hat type potential $V(\phi) = \frac{1}{2}\phi^2\left(m^2 - \lambda\phi^2\right)$.

which after a rescaling via (12), reads

$$\frac{\partial^2\tilde{\phi}}{\partial\tilde{x}^2} = \tilde{\phi}\left(1 - 2\tilde{\phi}^2\right) . \tag{28}$$

The potential (26) has only one minimum (maximum) for $\phi = 0$ at x_0. This zero implies a solution of the *lump* type. Moreover, the reflection symmetry of the potential permits two implicit solutions of the Bogomolny equation

$$\frac{d\tilde{\phi}}{d\tilde{x}} = \pm\tilde{\phi}\sqrt{\left(1 - \tilde{\phi}^2\right)} \tag{29}$$

which for $\tilde{\phi} > 0$ or $\tilde{\phi} < 0$ are

$$\tilde{x} - \tilde{x}_0 = \pm\int_{\tilde{\phi}(\tilde{x}_0)}^{\tilde{\phi}(\tilde{x})} \frac{d\tilde{\phi}}{\tilde{\phi}\sqrt{\left(1 - \tilde{\phi}^2\right)}} . \tag{30}$$

After integration and inversion we find

$$\tilde{\phi}(\tilde{x}) = \pm\text{sech}\left(\tilde{x} - \tilde{x}_0\right) , \tag{31}$$

again one for $\tilde{\phi} > 0$ and another for$\tilde{\phi} < 0$, see Fig. 6.

Now we can use the defining relation

$$\frac{dW}{d\tilde{\phi}} = \pm\frac{m^2}{\sqrt{\lambda}}\tilde{\phi}\sqrt{1 - \tilde{\phi}^2} , \tag{32}$$

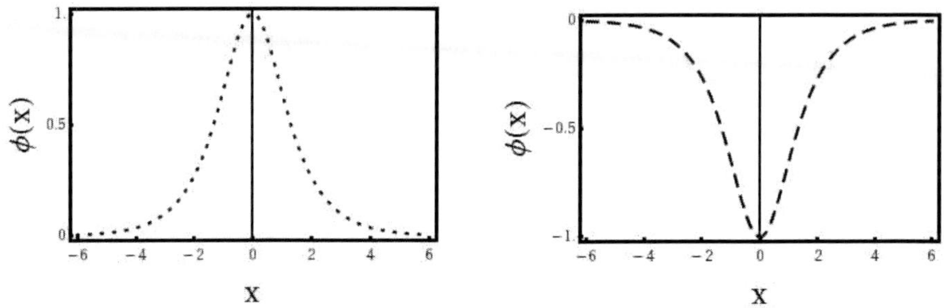

Figure 6. Soliton of the *lump* type, centered at the origin.

for the superpotential, which after integration leads to

$$W\left(\tilde{\phi}\right) = \mp\frac{m^2}{3\sqrt{\lambda}}\left(1-\tilde{\phi}^2\right)^{3/2} \tag{33}$$

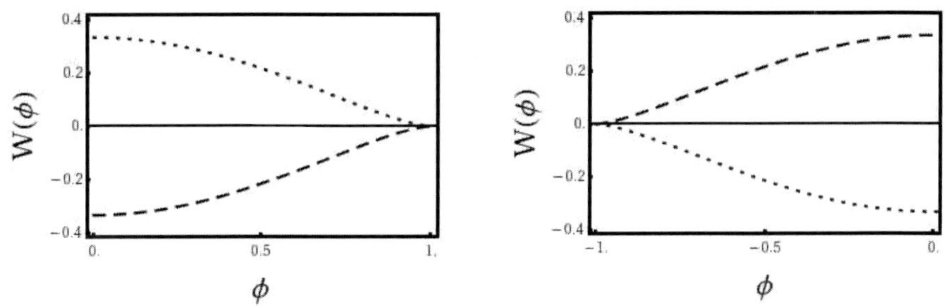

Figure 7. Higgs type superpotentials.

In Fig. 7 we see that the solution $\tilde{\phi}\,(\tilde{x})$ is positive or negative, depending on the range $[0, \pm 1]$ of the scalar field. The corresponding energy

$$\begin{aligned} E_{\text{lump}} &= \left|W\left(\tilde{\phi}_-\right) - W\left(\tilde{\phi}_0\right)\right| c^2 + \left|W\left(\tilde{\phi}_0\right) - W\left(\tilde{\phi}_+\right)\right| c^2 \\ &= 2\left|W\left(\tilde{\phi}_0\right) - W\left(\tilde{\phi}_+\right)\right| c^2 \\ &= 2\left|W(1) - W(0)\right| c^2 = \frac{2}{3}\frac{m^2}{\sqrt{\lambda}}c^2 , \end{aligned} \tag{34}$$

where $\phi_0 = \phi\,(x_0)$, cf. Fig. 8.

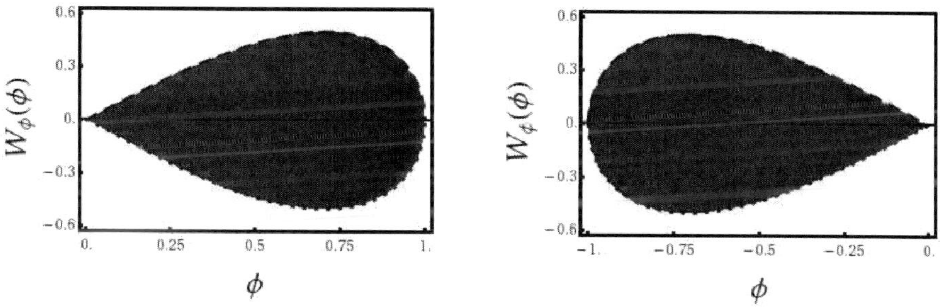

Figure 8. Energy of a lump calculated from $W_\phi\left(\phi\right)$.

5. Lane-Emden Type Potential

In the astrophysics of gaseous spheres [14, 22], like the Sun, for quite some time, the density has been modeled by a potential of the ϕ^6 type, cf. Refs. [21, 10].

In applying this to solitary waves, let us depart from the symmetric potential

$$V\left(\phi\right) = m^2\phi^2\left(1 - \chi\phi^4\right) = V\left(-\phi\right), \tag{35}$$

where m is a mass parameter and χ a coupling constant for the non-linear self-interaction, cf. Fig. 9.

The form of the potential depends essentially on the coupling constant χ: For $\chi \leq 0$ there is only one local minimum, whereas for $\chi > 0$ there are, additionally, two local maxima, all vacuum states. Obviously, for $\chi = 0$ the dynamics reduces to the linear Klein-Gordon equation, whereas $\chi < 0$ would imply imaginary mass terms as in the celebrated Higgs mechanism. Here, we will consider only the case $\chi > 0$.

Then the static KG equations takes the form

$$\frac{\partial^2\phi}{\partial x^2} = 2m^2\phi\left(1 - 3\chi\phi^4\right) . \tag{36}$$

which may lead, at least, to two types of solitonic solutions: Non-topological lump type solutions as well as topological ones such as kinks and antikinks.

Lump Type Solutions

The zeros $\phi^2 = 0$ and $\phi^4 = 1/\chi$ of the potential are degenerate, whereas the critical points are occurring for $\phi = 0$ and $\phi^4 = 1/3\chi$. Thus, there is only the

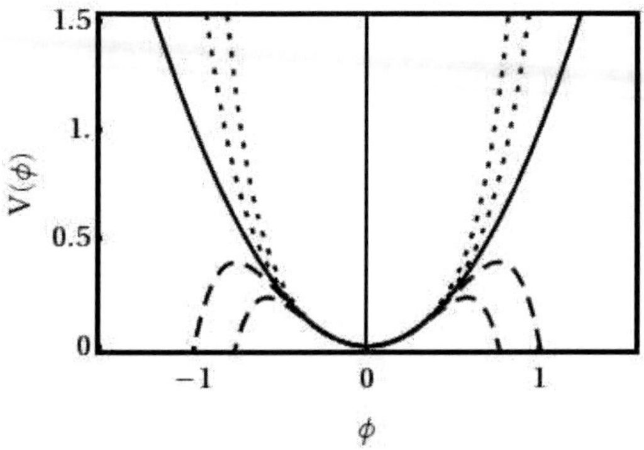

Figure 9. Lane-Emden potential $V(\phi) = m^2\phi^2\left(1 - \chi\phi^4\right)$. For $\chi = 0$, it reduces to quadratic potential (solid line), when $\chi < 0$ there arise parabolic curves (punctuated lines), whereas for $\chi > 0$ the potential has two maxima and one local minimum (dashed line).

point $\phi = 0$ which minimize the energy implying that the corresponding solution is non-topological of *lump* type centered at x_0. The Bogomolny equation now reads

$$\frac{d\phi}{dx} = \pm\sqrt{2}m\phi\sqrt{1 - \chi\phi^4}. \tag{37}$$

Since the potential is symmetric under reflection, there will be two solutions, one for $\phi > 0$ and another for $\phi < 0$. Moreover, in order that the integral remains reel, we require that $\chi\phi^4 \leq 1$. Then

$$x - x_0 = \mp\int_{\phi(x_0)}^{\phi(x)} \frac{d\phi}{\sqrt{2}m\phi\sqrt{1 - \chi\phi^4}} \tag{38}$$

and an inversion leads to the explicit solutions

$$\phi\left(x\right) = \pm\frac{1}{\chi^{1/4}}\left[\mathrm{sech}\left(2\sqrt{2}m\left(x - x_0\right)\right)\right]^{1/2}, \tag{39}$$

again one for $\phi > 0$ and another for $\phi < 0$, cf. Fig. 10.

Now we integrate once more the differential condition

$$\frac{dW}{d\phi} = \pm\sqrt{2}m\phi\sqrt{1 - \chi\phi^4} \tag{40}$$

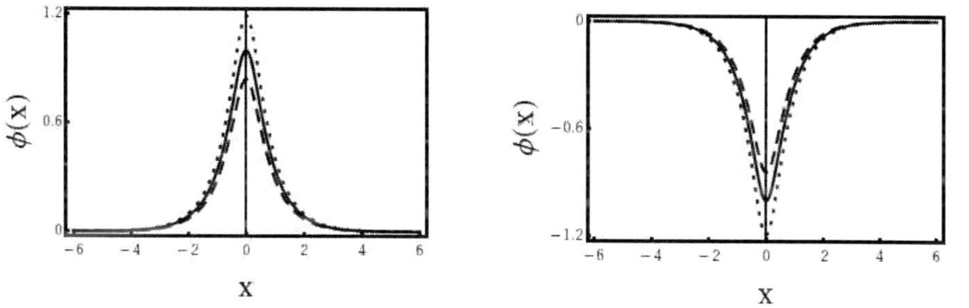

Figure 10. *Lump* type solution centered at $x_0 = 0$. Here, the value of $\chi = 0.5$ corresponds to punctuated line, $\chi = 1$ to the solid line and $\chi = 2$ to the dashed line.

in order to obtain the superpotential

$$W(\phi) = \mp \frac{\sqrt{2}}{4} m \left(\frac{\arcsin\left(\chi^{1/2}\phi^2\right)}{\chi^{1/2}} + \phi^2 \sqrt{1 - \chi\phi^4} \right). \tag{41}$$

cf. Fig. 11.

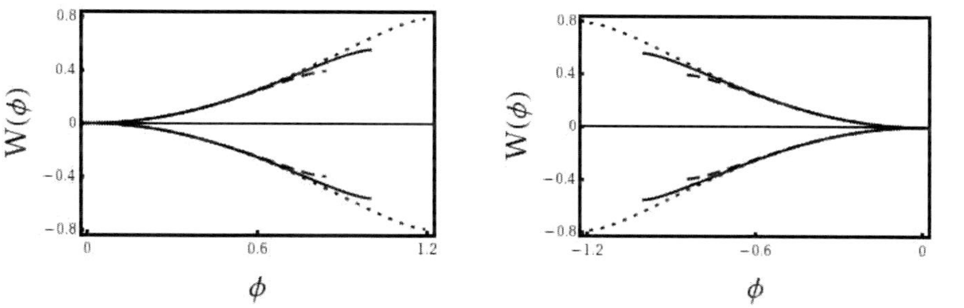

Figure 11. Superpotentials for the cases $\phi > 0$ and $\phi < 0$. Again, they are presented for $\chi = 0.5$ by a punctuated line, in the case $\chi = 1$ by solid line and for $\chi = 2$ by a dashed line.

The solution $\phi(x)$ is positive or negative, depending on the range

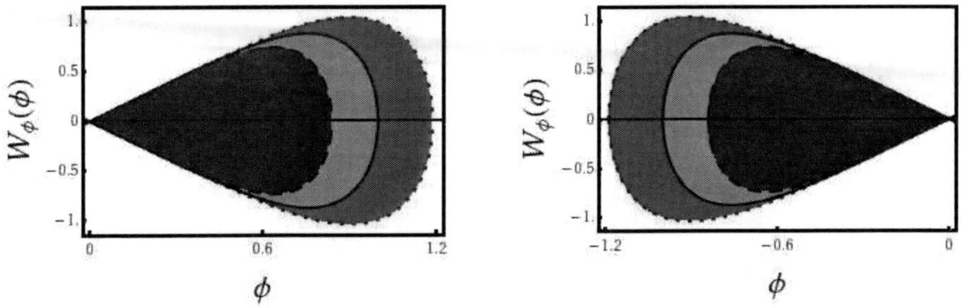

Figure 12. Energy calculated from $W_\phi(\phi)$ for $\phi > 0$ and $\phi < 0$. As before, we present superpotentials for $\chi = 0.5$ corresponding to the punctuated line, $\chi = 1$ to the solid line and $\chi = 2$ to the dashed line.

$\left[0, \pm 1/\chi^{1/4}\right]$. Therefore, the rest energy turns out to be

$$\begin{aligned} E_{\text{lump}} &= \left|W(\phi_-) - W(\phi_0)\right| + \left|W(\phi_0) - W(\phi_+)\right| \\ &= 2\left|W(\phi_0) - W(\phi_+)\right| \\ &= 2\left|W\left(\frac{1}{\chi^{1/4}}\right) - W(0)\right| = \frac{m}{2\sqrt{2}}\frac{\pi}{\chi^{1/2}}, \end{aligned} \tag{42}$$

see Fig. 12.

6. Outlook: Soliton Collisions

Recently, the so-called *'bullet cluster'* (catalog number 1E0657-558) provides indirect evidence [8] for the existence of DM, see also Ref. [17] for a dynamical simulation of the baryonic and DM components. In fact, for Zwicky the observed dynamics of galaxies in the Coma cluster provided the first clue in postulating already 1933 'missing matter' [27]. During a collision or merging of galaxy clusters with velocities up to 4700 km/s, the DM halos are displaced from the gas and stars and apparently pass each other in a soliton-like way. No annihilation will occur, if the halos are described by two coherent states of nonlinearly interacting scalar fields [5]. Although the observations in the bullet cluster [8] can be given a rather conventional interpretation [17], axionic BECs offer the possibility of lump-like DM structures [3], similarly to those of boson stars [19]. Since solitons are entities which carry a total mass, they behave effectively like particles which collide elastically, i.e. after leaving the interaction region where they may deform, they ultimately return to their original shapes and velocities. Since DM distribution in the Bullet Cluster seems not to be affected during merging, one may speculate about a solitary wave behavior of this component of galaxies and clusters.

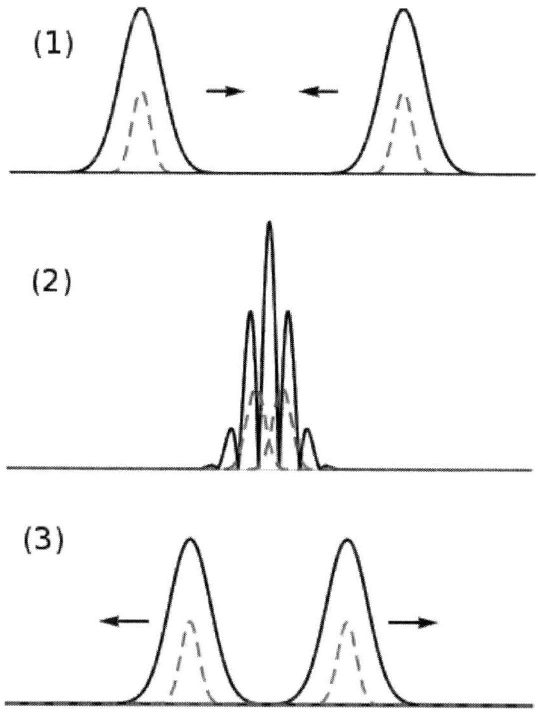

Figure 13. Collision of solitary waves similarly as that occurring in the Bullet cluster.

Fig. (13), adapted from Ref. [13], sketches the idealized distributions of hydrogen gas (drawn in red) and dark matter (drawn in black) for a galaxy cluster. The three parts are: In the wake of a collision (1); while colliding (2) when gas of both bullets is mixing while DM lumps get deformed but are essentially collisionless; and after collision (3), when two identical solitons emerge, except for possible change of their phases.

During the huge collision (Fig. 14) taking place in the Bullet Cluster, the hot gas [shown in red and detectable by X-rays] in each cluster is slowed and distorted by a drag force, similar to air resistance. A *bullet-shaped* cloud of gas forms in one of the clusters. In contrast, DM [shown in blue and only observable via micro-lensing] does not slow down by the impact because it does not interact directly with itself or the gas except through gravity. Therefore, the DM lumps from the two clusters move ahead of the hot gas, producing a separation of dark and normal visible matter.

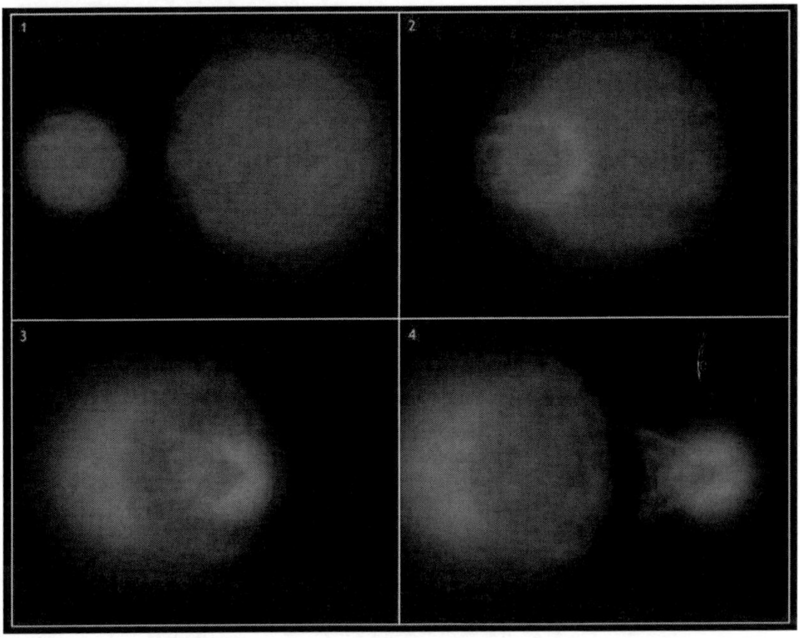

Figure 14. An artist's view of the collision taking place in the Bullet Cluster, as seen in four stages. (Credit: NASA/CXC/M. Weiss).

References

[1] A. T. Avelar, D. Bazeia and W. B. Cardoso: "Bright and dark solitons in a periodically attractive and expulsive potential with nonlinearities modulated in space and time, " *Nonlinear Anal. Real World Appl.* 11, 4269 (2010).

[2] A. T. Avelar, D. Bazeia, W. B. Cardoso and L. Losano: "Lump-like structures in scalar-field models," *Phys. Lett.* A374, 222-227 (2009).

[3] A. T. Avelar, D. Bazeia, L. Losano and R. Menezes: "New lump-like structures in scalar-field models," *Eur. Phys. J.* C55, 133-143 (2008).

[4] A. Barone, F. Esposito, C. Magee and A. Scott: "Theory and applications of the sine-Gordon equation", *Nuovo Cimento* 1, 227 (1971).

[5] A. Bernal and F. Siddhartha Guzman: "Scalar field dark matter: Head-on interaction between two structures," *Phys. Rev. D* 74, 103002 (2006).

[6] Y. Brihaye and T. Delsate: "Remarks on bell-shaped lumps: Stability and fermionic modes," *Phys. Rev. D* 78, 025014 (2008).

[7] D.-I. Choi: "Collision of gravitationally bound Bose-Einstein condensates," *Phys. Rev. A* 66, 063609 (2002).

[8] D. Clowe, M. Bradac, A. H. Gonzalez, M. Markevitch, S. W. Randall, C. Jones and D. Zaritsky: "A direct empirical proof of the existence of dark matter," *Astrophys. J.* 648, L109 (2006).

[9] R. K. Dodd, J. C. Eilbeck, J. D. Gibbon and H. C. Morris: *Solitons and Nonlinear Wave Equations* (Academic Press, London 1982).

[10] P. Dorey, K. Mersh, T. Romanczukiewicz and Y. Shnir: "Kink-antikink collisions in the ϕ^6 model," *Phys. Rev. Lett.* 107, 091602 (2011).

[11] P. Kumar and B. S. Thomas: "Internal modes of a soliton," *Phys. Rev. B* 37, 683 (1988).

[12] F. V. Kusmartsev: "On classification of solitons," *Phys. Scripta* 29, 7 (1984).

[13] J. W. Lee, S. Lim and D. Choi: *"BEC dark matter can explain collisions of galaxy clusters,"* arXiv:0805.3827 [hep-ph].

[14] F. K. Liu: "Polytropic gas spheres: An approximate analytic solution of the Lane-Emden equation," *Mon. Not. Roy. Astron. Soc.* 281, 1197 (1996).

[15] A. Mahdavi, H. Hoekstra, A. Babul, D. Balam and P. Capak: "A dark core in Abell 520," *Astrophys. J.* 668, 806 (2007).

[16] N. Manton and P. Sutcliffe: *Topological Solitons* (Cambridge University Press, Cambridge 2004).

[17] C. Mastropietro and A. Burkert: "Simulating the bullet cluster," *MNRS* 389, 967-988 (2008).

[18] J. Mielczarek, T. Stachowiak and M. Szydlowski: "Vortex in axion condensate as a dark matter halo," *Int. J. Mod. Phys.* D19, 1843-1855 (2010).

[19] E.W. Mielke and F. E. Schunck: "Boson stars: Alternatives to primordial black holes?", *Nucl. Phys. B* 564, 185 (2000).

[20] E. W. Mielke and J. A. Vélez Pérez: "Axion condensate as a model for dark matter halos," *Phys. Lett. B* 671, 174-178 (2009).

[21] E. W. Mielke, and J. A. Vélez Pérez: "Toroidal halos in a nontopological soliton model of dark matter", *Phys. Rev. D* 75, 043504 (2007).

[22] K. Parand, M. Dehghan, A. R. Rezaei and S. M. Ghaderi: "An approximation algorithm for the solution of the nonlinear Lane-Emden type equations arising in astrophysics using Hermite functions collocation method," *Comput. Phys. Commun.* 181, 1096-1108 (2010).

[23] R. Rajaraman: *Solitons and Instantons* (North-Holland, Amsterdam 1982).

[24] J. Ruostekoski: "Stable particlelike solitons with multiply-quantized vortex lines in Bose-Einstein condensates," *Phys. Rev. A* 70, 041601 (2004).

[25] R. Teisseyre, M. Takeo and E. Majewski: ed., *Earthquake Source Asymmetry, Structural Media and Rotation Effects* (Springer, New York 2006).

[26] N. J. Zabusky and M. D. Kruskal: "Interaction of 'Solitons' in a collisionless plasma and the recurrence of initial states," *Phys. Rev. Lett.* 15, 240-243 (1965).

[27] F. Zwicky: "Spectral displacement of extra galactic nebulae," *Helv. Phys. Acta* 6, 110 (1933).

INDEX

A

B

C

D

E

F

G

H

I

J

K

L

M

N

O

P

T

U

V

W

Y